During its five-billion-year history, Earth has been hit countless times by asteroids and meteorites. Over 150 crater-producing events have been identified, and this book describes 139 sites worldwide at which evidence of the inmpacts can be found. They range in age from recent craters formed this century to the highly eroded billion-year-old ancient craters. Some are spectacular to visit, such as the Barringer Crater in Arizona, the ring-shaped mountains of Gosses Bluff, Australia, and the huge crater at Ries in Germany. For each site there is a summary table giving location, size, age and present condition. Maps are included where necessary. The author has visited many of the sites and his photographs enrich this thorough survey. Meteorite craters are fascinating to visit, so the descriptions include guidance about access and suggested itineraries for the large structures.

Meteorite craters and impact structures of the Earth

This book is dedicated to crater-exploring pioneers Carolyn and Eugene Shoemaker, shown here making notes about meteorite craters while traveling in the Great Sandy Desert of Australia.

Meteorite craters and impact structures of the Earth

Paul Hodge

University of Washington

CAMBRIDGE UNIVERSITY PRESS
Cambridge, New York, Melbourne, Madrid, Cape Town, Singapore,
São Paulo, Delhi, Dubai, Tokyo

Cambridge University Press
The Edinburgh Building, Cambridge CB2 8RU, UK

Published in the United States of America by Cambridge University Press, New York

www.cambridge.org
Information on this title: www.cambridge.org/9780521126045

First published 1994
This digitally printed version 2009

A catalogue record for this publication is available from the British Library

Library of Congress Cataloguing in Publication data

Hodge, Paul W.
Meteorite craters and impact structures of the earth / Paul Hodge.
p. cm.
ISBN 0 521 36092 7
1. Meteorite craters – Handbooks, manuals, etc. I. Title.
QB755.H63 1994
551.3′97–dc20 93-21350 CIP

ISBN 978-0-521-36092-0 Hardback
ISBN 978-0-521-12604-5 Paperback

Contents

6 Impact structures of Europe

7 Impact structures of Africa

8 Impact structures of Asia

1 Introduction

The results of more than 150 crater-producing events have been identified on the Earth's land mass. These range from recent craters, formed before our eyes, as at Sikhote Alin, to ancient craters, eroded almost away and a billion years old, as at Acraman. Some are nicely preserved and spectacular places to visit, as at the Barringer Crater, and others are completely invisible, buried beneath hundreds of meters of sediments, as at Des Plaines. They range in diameter from a few meters, as at Henbury, to over 100 km, as at Vredefort. Whether large or small, young or old, clear or obscure, all of the structures are important records of how astronomical objects have continued to have an effect on the events that made up the history of the Earth.

How this book is organized

The following chapters of this book provide descriptions of the craters and impact structures, organized with a separate chapter for each continent. Within each chapter, the individual objects are described in alphabetical order.

Some difficulty of separation arises for objects that lie near the ill-defined boundary between Europe and Asia. This was not so serious a problem in the past, when the Soviet Union, which hosted most Asian impact structures, could be used as a delineating unit. However, with the dissolution of the Union, the craters lie in various political divisions, and so I have adopted the continental dividing line to separate them. Following tradition, I have used the Ural Mountains as the primary boundary between Europe and Asia, with the Ural River, the Caspian Sea and the Turkish and Iranian northern borders as the southern extension of the line.

For each crater and impact structure there is an initial summary table that lists its location, including its latitude and longitude, its size, its age (or known limits on its age) and its present condition. Positions and diameters were adopted (except for very newly recognized objects and for a few cases where there were some typographical errors) from the authoritative list compiled in 1991 by R. A. F. Grieve (*Meteoritics*, **26**, 175–194). In some cases, different sources give different values for the sizes of craters. This is usually the result of different choices having been made regarding the definition of the outer boundaries of structure, a particularly awkward problem for very large, complex, eroded cases. I have chosen to follow Grieve's decisions about sizes.

The descriptions of the craters and impact structures are based on both published accounts and personal experience. I have visited many of the objects, photographed them, and in some cases have done some scientific research, in particular on the meteoritic and related microscopic objects that are found in the soil surrounding the younger craters. But much of the description has been based on the accounts of those scientists who have most thoroughly explored the objects and most completely described them in print.

Because many meteorite craters are fascinating places to visit, providing the only chance for most of us to see and study and walk on real astronomical phenomena, I also provide some guidance about access. Where possible, this is based on personal experience, but for many I have had to rely on published accounts and detailed maps. As times change, many of the more remote sites are becoming increasingly accessible and so a prospective visitor would be wise to make up-to-date and local inquiries before planning an expedition to a crater locale. I have not visited any of the Asian and East European structures and therefore refrain from giving advice on how to reach most of them.

At the end of each description one or two references to published papers are given. This is a selected sample for the better-studied craters. More complete reference lists are given, for example, by M. J. Grolier (*NASA Technical Memorandum 87567*, 1985). For some craters, such as the Snelling Crater in Australia, no published account is yet available. For many objects formerly located in the Soviet Union easily-accessible publications are also unavailable, explaining why many of them receive very limited coverage in this book.

How complete is this book?

The objects described in this book include all certain and probable impact structures known to me in 1992. I have used Grieve's list, cited above, as my criterion, with only eight objects added from the more recent literature. Many other objects around the world have been proposed by various scientists, and some of these will probably be admitted to 'probable' or 'certain' status with further study.

Many, however, will not. The meteorite-crater literature is rife with false alarms. Because really good criteria for an impact origin have only recently been developed, the subject has seen wild swings back and forth in the past, between claims that obvious volcanic features are meteorite craters to claims that obvious meteorite craters, such as the Barringer and Wolfe Creek craters (where fragments of the responsible meteorite are abundant), are not. As explained below, evidence for the production of high pressure shock effects now provides a reasonably reliable method for recognizing the real thing.

Meteorite craters versus impact structures

Readers of the above paragraphs and those who paid attention to the title of this book many wonder why 'meteorite craters' and 'impact structures' are mentioned and discussed as if they were different kinds of object. They really are not different in origin, but workers in this field usually distinguish them in this way in order to separate the relatively fresh, uneroded 'meteorite craters' that still have their complete craterform anatomy from the older, eroded and nearly-unrecognizable 'impact structures', of which we only see the skeletons. In some cases, as for Gosses Bluff, a casual glance might suggest that what we see is a complete crater, but a careful field study shows that the 'rim' of the apparent crater is really a ridge of rock hardened by the original impact, but located inside and deep below the original crater, which was far larger and has long since been eroded away.

The cratering mechanism

Impact cratering is a primary means for forming the surface characteristics of many objects in the solar system – the Moon, Mercury, much of Mars and its satellites, most of the moons of the major planets, the asteroids, and even interplanetary spacecraft. Study of the cratering mechanism thus has widespread relevance and much progress in understanding it has been made in recent years. There are several excellent books on the subject, of which I mention here only a sample:

Roddy, D. J., Pepin, R. O. and Merrill, R. B., 1977, *Impact and Explosion Cratering.* Pergamon Press, New York.

French, B. M. and Short, N. M., 1968, *Shock Metamorphism of Natural Materials.* Mono Book Corp., Baltimore.

Rim of the Barringer crater in Arizona, showing the characteristic upturned layers at the top of the steep crater walls.

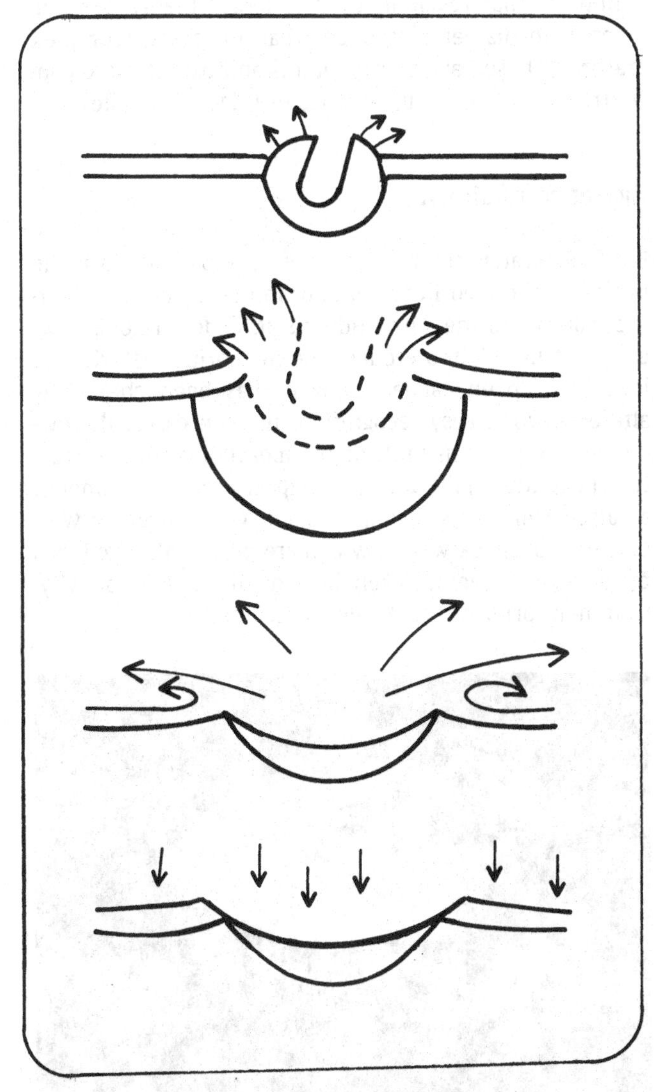

For a relatively small crater like the Barringer crater, the formation can be represented by this sequence of events. Almost all of the meteorite and the rocks at the point of contact are explosively destroyed. A shock wave radiates out through the rocks and debris is thrown out into the atmosphere. The rock layers are upturned and overturned at the rim and the fragments of shattered rocks form a lens below the depression, which then fills in with the material that was ejected.

A very brief summary of the events involved in meteoritic impact on the Earth is given here:

1. When a large meteorite encounters the Earth its passage through the atmosphere is very brief (unless the angle of incidence is shallow), only taking a few seconds. Sometimes the collision with the atmosphere breaks the projectile into several fragments.
2. A shock wave is formed at the prow of the meteorite, where there is a layer of highly-compressed air together with a small amount of melted and evaporated meteorite material.

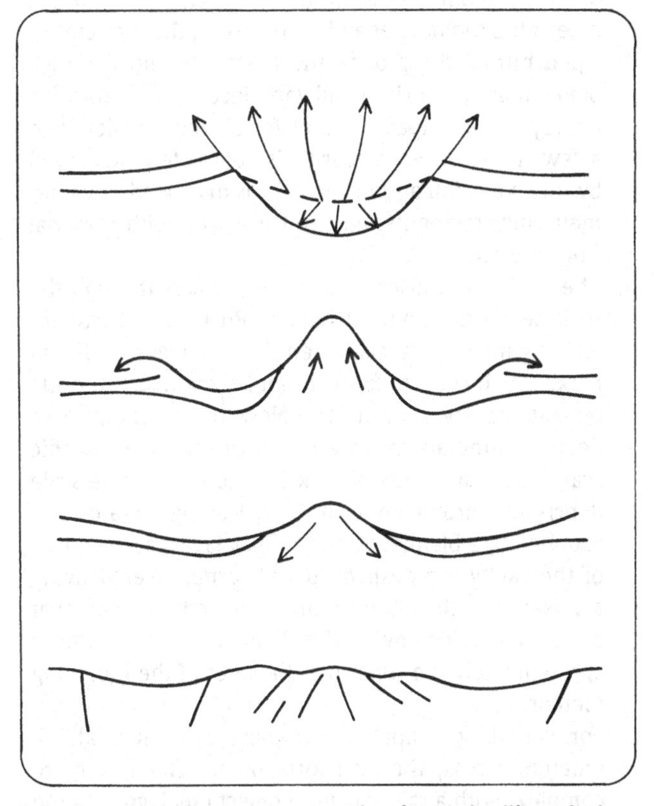

Large impact structures are more complex. A central uplift results from the rebounding rocks, which undergo compression and ejection. In addition to the large amount of rock fragments, vapor and liquid, material flows out along the surface in waves. The final form results from the collapse of the central peak and downfaulting in the outer parts, forming a ring or rings.

Small craters, like this one at Henbury in Australia, sometimes still retain some of the impacting meteorite. Meteorites were excavated from this crater in 1932.

3. For very large impactors, the frontal layer of compressed gas is capable of melting and ejecting blobs of rock before the solid object touches the Earth's surface. These blobs are the tektites and are often found hun-

dreds of kilometers away from the crater, as in the case of the Moldavites and their parent, the Ries crater.

4. Upon hitting the ground the meteorite and the rock below it are pulverized and vaporized by the explosive energy of the impact. Except for objects smaller than a few meters, the meteorite is completely destroyed by the explosion and its material is dispersed as vapor, melt, and fragments, much of it mixed with material from the target.
5. The explosion causes shock waves to pass through the rock below the point of impact. Rocks are shattered, forming impact breccia. Often there is a layer of liquid rock, which cools to form an impact melt, superficially resembling lava. Even far below the impact, rocks develop minerals that are distorted on a microscopic scale. Large amounts of rock fragments and fine-scale debris are thrown from the site, leaving a crater surrounded by a blanket of ejecta. Rock layers at the edges of the cavity are pushed up and turned over, leaving a raised rim that is overlain with beds of rock that are now up side-down. The diameter of the crater is approximately ten times the diameter of the incoming meteorite.
6. For very large impacts, forming craters several kilometers across, the final form of the crater is more complex, with a raised central object (analogous to the central peaks of lunar and Martian impact craters) and a series of concentric rings in the outer regions, made up of alternating uplift areas and depressions. Because such large cratering events occur very rarely (fortunately), terrestrial examples are all highly eroded. Often all that remains is the central uplift and the circular zone of shocked and shattered basement rock.

Types of crater

Meteoriticists often divide impact craters on the Earth into three types, according to the size of the impact. For small meteorites, up to sizes of a few meters, the energy of the impact is usually insufficient to produce a true explosion, and the meteorite plows into the ground, making what can be called a **'dug crater'**. This is usually several times the size of the meteorite, but for small meteorites hitting soil rather than rock, it can be only as large as the meteorite.

Larger impacting bodies produce an explosion, with shock waves passing through the target, leaving the many features described above. For craters that are tens of meters up to a few kilometers in diameter, the result can be called a **'simple crater'**. It is symmetrical and perfectly bowl-shaped.

Impacts that result in cavities larger than a few kilometers in diameter produce what are called **'complex craters'**. These have central uplifts and two or more concentric rings making up a series of ridges and valleys.

Evidence of shock

For fresh craters the best proof of an impact origin is the presence of meteorites around or in the crater or meteoritic material in the surrounding soil. Older craters, however, are usually too eroded; the meteoritic material has long since been carried away or has been chemically altered beyond easy recognition. In such cases, the evidence of impact must rely on the morphology of the structure (a central uplift, a lens of impact breccia, symmetry, a raised rim, pools of impact melt, etc.), together with evidence of shock waves, which are not usually produced by terrestrial events, even by violent volcanic activity. Common forms of shock effects include:

A large crater is formed with such high pressures that shatter-coning can occur in the target rocks. This shatter cone was retrieved from the Sierra Madera structure in Texas.

1. Shatter cones. When strong shock waves pass through appropriate types of rock, they produce weakened zones radiating out from the direction of shock, resulting in ridged cone-shaped rock fragments when the shocked layers are bared by erosion. Much of the large spurt of progress in recognizing impact structures that occurred in the 1960s was the result of the discovery of shatter cones at many previously dubious sites.
2. Shock lamellae and linear features in quartz. When subjected to strong shocks, crystals of quartz and other minerals show microscopic parallel linear patterns, called shock lamallae. Sometimes there are several systems at different angles. These kinds of features are

common in lunar rocks, for which impacts have been the primary geological process.

3. Shock-induced minerals. Certain minerals, such as coesite (a high-pressure phase of silica that is formed at 450–800 °C and 38 000 atmospheres) and stishovite (a high-density form of silica formed at 130 000 atmospheres), are strong confirmation of an impact origin for a feature.

Key to Maps

thin line	road
thick line	outline of crater or structure
—/	faults
wavy lines	stream or river
- - - - -	rock type boundaries
hatched line	political boundaries

2

Impact structures of the United States

Barringer

Arizona

Lat/Long: N35° 2′, W111° 1′
Diameter: 1.2 km
Age: 0.049 Ma
Condition: Fresh

The Barringer Crater, also known as the Arizona Crater or Meteor Crater, is the best known impact structure in North America. At least two excellent books have been written about it (see below) and it played an important role in the development of the history of the field, being the first universally-recognized example of a meteorite crater. The story of this history, in the first three decades of the twentieth century, is fascinating, involving an 'outsider' with a radical theory and a powerful establishment figure who held tenaciously to his conservative views in the face of what (to us now, at least) seem to be unarguable facts. Daniel Moreau Barringer, a mining entrepreneur, spent much of his life and fortune attempting to convince the scientific community that the crater (locally known as 'Coon Butte') must have had an impact origin. The geology community, led by the chief geologist of the US Geological Survey, Grove Karl Gilbert, insisted that the crater was a volcanic explosive feature and that the existence of meteoritic fragments around it was merely a coincidence. The issue was not really resolved until about 1930, when enough studies had finally convinced enough scientists that a meteoritic origin was the most likely explanation of the facts. Much of the resistance to the acceptance of this idea had been caused by Barringer's conviction that a huge mass of iron–nickel lay beneath the crater floor. When such an object was not found, the entire hypothesis was faulted, even when the astronomer F. R. Moulton showed the such an object couldn't withstand an impact whole and would be destroyed and dispersed.

The Barringer Crater has the outline of an imperfect circle, nearly polygonal. Its rim rises 45 m above the surrounding, very flat, high Arizona desert, and is nearly uniformly high around the entire perimeter. The flat floor of the crater (which at some time or times, according to sediments, was lake-filled) lies 100 m below the surroundings. Its inner slopes are steep and rugged, while the outer slopes are gentle, marked by large and small deposits of debris from the cratering process. The nearly flat-lying beds of sedimentary rock are upturned radially in the rim, and in some areas it is possible to see that they have

BARRINGER: The east rim of the Barringer crater. Edges of uptilted strata are visible in the upper half of the inner wall.

BARRINGER: The floor of the Barringer crater is 145 m below the rim. It is filled with Recent and Pleistocene sediments.

BARRINGER: The south rim of the Barringer crater. Blocks of ejecta and upturned beds are visible. The rim trail is at left.

been overturned near the edge, with the normal vertical sequence reversed. Although made up of fragments and somewhat mixed, these beds form a nicely overturned flap surrounding the crater.

Thousands of meteorite fragments have been recovered from the Barringer Crater, mostly from the outer rim, but also from as far as 7 km from the crater. Over 10 000 ponderable meteorites have been recovered and many more micrometeorite-sized specimens have been found. One study of the soil surrounding the crater showed that meteoritic dust pervades the soil out to distances of nearly 10 km from the crater and that the total mass in this form is on the order of 12 000 metric tons. Compared to the 30 or so tons in the form of large meteorites, this is by far the largest form for the meteoritic matter, and it corresponds fairly closely to estimates of the total mass necessary to produce a crater of this size.

Other forms of impact materials include large numbers of shale balls, some meteoritic-laced impact glass and various forms of shock-induced minerals, including small diamonds, coesite and stishovite.

Anyone interested in impact craters should visit the Barringer Crater (the other essential visit is to the Ries and its remarkable museum in Nordlingen). It is easy to visit (see below) and is so well preserved that many of the features of moderate-sized impact structures are displayed well. At the time of writing, visitors are generally not permitted to climb down into the crater, but a good trail exists around the rim, from which many important features are visible. A clockwise itinerary follows. But first it is useful to identify the various rock formations that exist in the region. Before the meteorite landed, the geology of the area was relatively simple. Basically, there were nearly horizontal layers of Mesozoic and Paleozoic sedimentary rock. Leaving out detail, the following table lists the principle layers and their approximate thicknesses near the present position of the crater:

Formation	Age	Thickness (m)
Alluvium	Recent and Pleistocene	Thin, variable
Moenkopi (sandstone)	Triassic	Thin, variable
Kaibab (limestone and dolomite)	Permian	90
Coconino (sandstone)	Permian	270

The main outcrops near and at the crater rim are of the Kaibab and Moenkopi formations, while the Coconino sandstone is exposed in the inner crater walls.

BARRINGER: Beds of sandstone on the western rim of the Barringer crater, lying vertically (foreground) and at an intermediate angle (background).

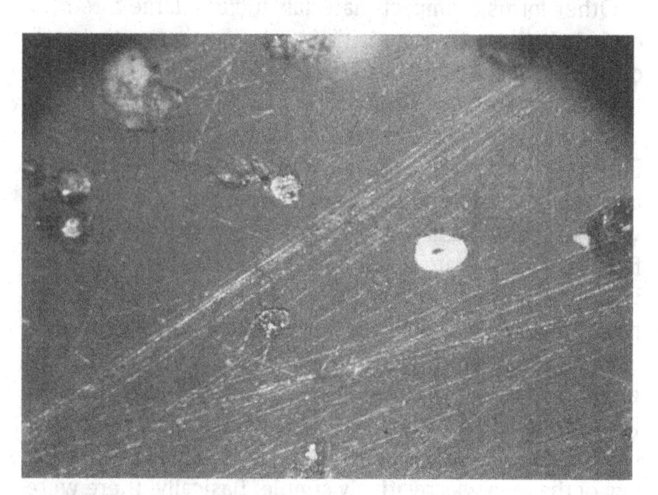

BARRINGER: Microscopic meteoritic and iron oxide particles from the soil, collected 10 km from the crater.

The following is an itinerary for a walk around the crater rim on the Rim Trail, proceeding clockwise:

Distance from museum (km)	Features to note
0	There is a path to the left from the museum. A few meters from the building, just before a branching path leads to the left up the ridge to a lookout platform, there is an exposure of upturned bedrock. The uppermost beds are of red Moenkopi sandstone and they are sharply tilted and, in some places, overturned.
0.1	An excellent view of the rim and floor of the crater is provided from the lookout platform at the top of the path. The yellowish blocks surrounding the platform are Kaibab limestone, forming a blanket of debris that occupies much of the outer flanks of the crater. The stratification of the Kaibab formation is found to be inverted here, compared to the undisturbed strata, demonstrating that the cratering process overturned the layers of rock as it lay down the debris. The rim trail proceeds eastward from the lookout.
0.4	The rim trail passes through a field of blocks of Kaibab limestone and dolomite. Note the immense size of those on the outer flanks of the rim, which were thrown out of the crater cavity whole.
0.9	Small exposures of Moenkopi sandstone are found on the inside of the rim, related to small nearly radial faults.
1.2	The trail passes near debris of Coconino sandstone, which lies along much of the southern outer rim. Trees, absent elsewhere in the crater area, grow here.
1.5	Here is the site of the drilling carried out at the rim in 1920–22, when it was hoped that a large meteoritic mass might exist beneath the south rim, assuming that the meteorite entered from the north. Timbers and excavations remain. The cut in the rim shows exposures of breccia related to the faulting that formed this cut.
1.6	The trail divides beyond the drilling site, with one section proceeding along the inner rim and the other on the outer. The inner branch descends a few meters and passes through outcrops of red sandstone.
2.0	Several derelict buildings are encountered on the southern flank of the crater.
2.2	The highest point on the rim is reached.
2.8	The trail descends onto the western outer slope, where it threads its way through huge blocks of Kaibab formation.
3.1	The picturesque ruins of Barringer's

Distance from museum (km)	Features to note
	original museum are passed. From here the old mule trail descends into the crater. A spur trail leads out onto flat shelves of red sandstone, providing a superlative view down into the crater.
3.4	Trail passes above outcrops of uptilted Moenkopi sandstone.
3.7	Return to the museum.

Access: One of the easiest meteorite craters in North America to visit, the Barringer Crater is located about 7 km south of US 66 in northern Arizona, east of Flagstaff. A paved access road leads to an excellent museum, a shop, and a snack bar at the rim of the crater. The crater is privately owned and an admission fee is charged.

References

Hoyt, W. G., 1987, *Coon Mountain Controversies.* University of Arizona Press, Tucson, AZ, 443 pp.

Nininger, H. H., 1956, *Arizona's Meteorite Crater.* American Meteorite Laboratory, Denver, CO, 232 pp.

Shoemaker, E. M., 1960, Penetration mechanics of high velocity meteorites, illustrated by Meteor Crater, Arizona *Report of the International Geological Congress, XXI Session*, Part XVII, pp. 418–434.

Shoemaker, E. M. and Kieffer, S. W., 1974, *Guidebook to the Geology of Meteor Crater, Arizona.* Publication 17, Center for Meteorite Studies, Tempe, Arizona, 66 pp.

Sutton, S. R., 1985, Thermoluminescence measurements on shock-metamorphosed sandstone and dolomite from Meteor Crater, Arizona. 2. Thermoluminescence age of Meteor Crater. *J. Geophys. Res.*, *90*, 3690–3700.

Beaverhead

Montana

Lat/Long: N45° 0′, W113° 0′
Diameter: 15 km
Age: ~600 Ma
Condition: Eroded, partly exposed

The Beaverhead impact structure was only recognized in 1990, having been identified on the basis of the presence of shatter cones in outcrops of a geologically complex area. The shatter cones were found at Island Butte, a mountainous structure at the Montana–Idaho border next to the Continental Divide. The area is marked by many fault zones and includes rocks from the Precambrian to the Cenozoic. The shatter cones are found over an area of approximately 17 km radius, but they all show an identical orientation (with apexes at 90°), and thus it is inferred that the original crater may have been as large as 60 km in diameter. The cones are found in outcrops of Proterozoic sandstones and are absent in adjacent Mississippian rocks, indicating that the impact must have occurred during the late Precambrian or early Paleozoic. Some samples of the sandstones show veins that may be melt flows from impact melting.

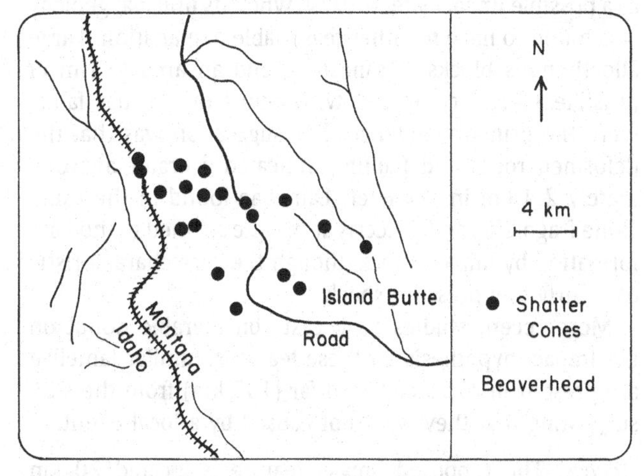

BEAVERHEAD: Sketch map of the Island Butte area, showing locations of Beaverhead shatter cones (after Hargraves *et al.*).

Access: The shatter cones are spread out over an area of ~100 km² in southwest Montana approximately 50 km west of Interstate 15 and 125 km southwest of Butte, Montana. A mountain road passes through the center of the shattercone field.

References

Hargraves, R. B., Cullicott, C. E., Deffeyes, K. S., Hougen, S., Christiansen, P. P. and Fiske, P. S., 1990, Shatter cones and shocked rocks in Southwestern Montana: the Beaverhead impact structure. *Geology, 18*, 832–834.

Koeberl, C. and Fiske, P. S., 1991, Beaverhead impact structure, Montana: geochemistry of impactites and country rock samples. *Meteoritics, 26*, 358–359.

Bee Bluff

Texas

Lat/Long: N29° 2′, W 99° 51′
Diameter: 2.4 km
Age: <40 Ma
Condition: Partly exposed

This entry is included in the book as an example of the numerous cases of proposed impact structures that have been considered doubtful and that have recently been omitted from authoritative lists of impact structures. The Bee Bluff structure (also called Uvalde) was first proposed as a possible impact site in 1979, when its unusual geology was found to have no other reasonable explanation. Large allocthonous blocks of sandstone and an arcuate rim of possible ejecta, combined with numerous thrust faults were the primary evidence. The suggestion was that the deformed rocks and faulting indicated a crater approximately 2.4 km in diameter. Lamellae found in the sandstone fragments and breccias at the site seemed to indicate formation by impact shock, though they are characteristic of a fairly low pressure shock.

More recent studies have cast considerable doubt on the impact hypothesis for these features. Similar lamellae are present in rocks collected far (135 km) from the site, suggesting that they were not caused by a local event.

Access: The proposed impact feature is located 20 km south of Uvalde, Texas. US highway 83 passes straight through it, bisecting the arc of proposed ejecta material. The east side of the proposed impact rim nearly coincides with a bank of the Nueces River.

References

Sharpton, V. L. and Nielsen, D. C., 1988, Is the Bee Bluff structure in S. Texas an impact crater? *Lunar Planet. Sci., 19*, 1065–1066.

Wilson, W. F. and Wilson, D. H., 1979, Remnants of a probable Tertiary impact crater in South Texas. *Geology, 7*, 144–146.

Calvin

Michigan

Lat/Long: N42°, W86°
Diameter: 7 km
Age: 460 Ma
Condition: Buried

The Calvin Structure was delineated by cores from 107 wells drilled in Cass County, southwestern Michigan. The surface expression of the structure is minimal; a slightly higher topography at the center of the feature shows up because of its influence on the drainage in the area. The entire structure is deep below the present surface, which is underlain by ~100 m of glacial drift that sits on top of 1300 m of Paleozoic strata.

The drilling shows the presence of a central uplift about 400 m high, surrounded by an annular depression about 1 km wide. An anticlinal rim zone, consisting of an annulus 1.5 km wide and several meters high, surrounds this depression. Microbreccia exists in the rocks below the supposed surface of the crater-shaped structure and the fact that it is found in layers of widely differing ages indicates that it was probably formed by an impact.

Shock-metamorphosed quartz grains were identified in 1992 and possible iron-melt spherules were found in the outside walls of the structure.

Access: There is nothing to see at the surface of the Calvin Structure. It is located beneath a portion of Michigan that is easy to reach, however. Adamsville is near one edge of the structure and Calvin Center (12 km by road northeast of Adamsville) is near its center. A chain of six lakes, from Chain Lake to Curtis Lake, lies partly above the north portion of the structure.

References

Milstein, R., 1988, Impact origin of the Calvin 28 cryptoexplosion disturbance, Cass County, Michigan. *Lunar Planet. Sci., 19*, 778–779.

Milstein, R., 1988, The Calvin 28 structure: evidence for impact origin. *Can. J. Earth Sci., 25*, 1524–1530.

Crooked Creek
Missouri

Lat/Long: N37° 50′, W91° 23′
Diameter: 7 km
Age: 320 Ma
Condition: Partly exposed

The Crooked Creek structure is one of the many circular disturbed areas in the midwestern US that were originally thought to be cryptovolcanic in nature. In the early 1960s, even after shatter cones were found there (and before shatter cones were considered definitive evidence of impact-induced shock), it was still considered non-meteoritic.

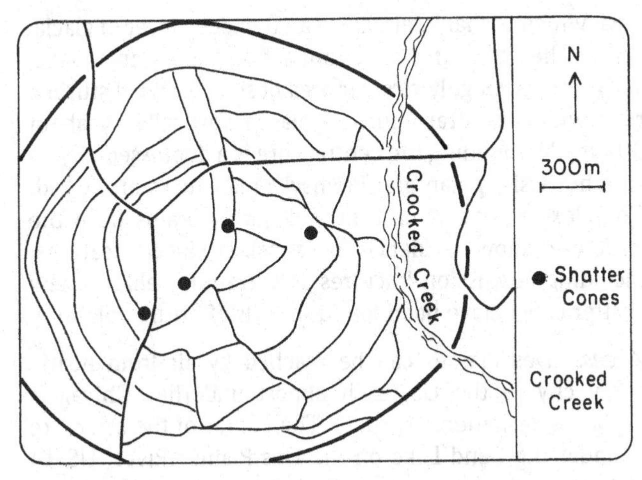

CROOKED CREEK: The Crooked Creek structure, with the locations of faults indicated (after Kiilsgaard *et al.*).

The structure is underlain by Cambrian and Ordovician sedimentary rocks, mostly dolomites, which in the surroundings in this part of Missouri, lie nearly flat and undisturbed. The central area consists mostly of Potosi Dolomite, in which good examples of shatter cones have been found exposed in road cuts. Several faults cross the structure and it is bounded by a series of normal faults. There is an encircling graben of young (Ordovician) sediments and a central uplifted zone of older (Cambrian) rocks. The central strata are approximately 300 m above their normal position for the area. The uplifted rocks are highly brecciated. Mining of several minerals has occurred in the central area, but the mineralization appears to be more recent than the impact event.

Access: The Crooked Creek structure is 13 km southeast of Rolla, Missouri and 4 km directly south of Steelville. It is trisected by three roads, which intersect about 300 m northwest of the center of the central uplift.

References

Kiilsgaard, T. H., Heyl, A. V. and Brock, M. R., 1963, The Crooked Creek disturbance, Southeast Missouri. *US Geol. Surv. Prof. Pap.*, No. 450-E, 14–19.

Decaturville
Missouri

Lat/Long: N37° 54′, W92° 43′
Diameter: 6 km
Age: <300 Ma
Condition: Partly exposed

The Decaturville structure attracted early interest because of its unusual geology and because of the presence of mineral deposits, especially of lead, zinc and iron sulfide. The first geological interpretations were varied and included identification of it as a Cambrian mud volcano. The structure is ringed by a nearly circular fault, which surrounds a depressed ring-shaped zone and a central uplift. Good rock outcrops are found at the center of the feature, which has the structure of a polygonal uplift, surrounded by complex, disturbed and highly-faulted sedimentary rocks, mostly dolomites and limestone. Surface mapping and drill cores show overturned and overthrust sequences.

By the 1970s it was realized that the Decaturville Structure has many of the characteristics of the cryptovolcanic features that were turning out to be impact structures. A suggestion made in 1936 that it might instead be the result of meteorite impact was therefore brought forward

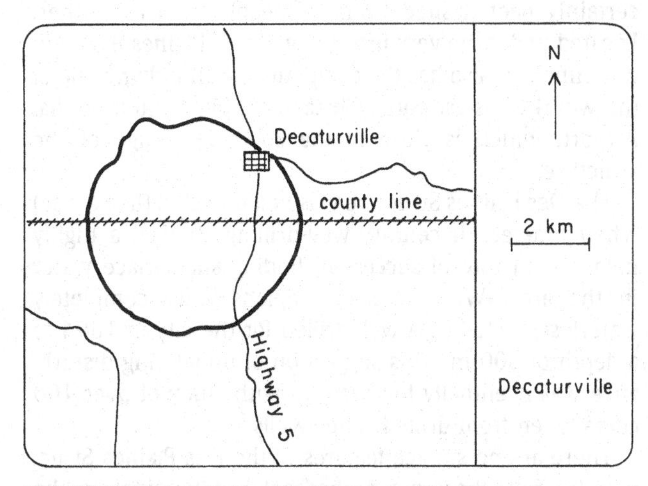

DECATURVILLE: Location map for the Decaturville structure.

again and eventually confirmed by the discovery of impact breccias and various shock-deformation features, including shatter cones and planar features in quartz. The shatter cones are common in the dolomites at the very center.

Access: The Decaturville structure lies 13 km south of the town of Camdenton, on the Ozark Plateau east of the Lake of the Ozarks. Highway 5 from Camdenton passes through the eastern side of the structure and the village of Decaturville lies on the northeast portion of the outer ring.

References

Offield, T. W. and Pohn, H. A., 1977, Deformation at the Decaturville impact structure, Missouri, in *Impact and Explosion Cratering*, ed. D. J. Roddy, R. O. Pepin, and R. B. Merrill. Pergamon, Oxford.

Offield, T. W. and Pohn, H. A., 1979, Geology of the Decaturville impact structure, Missouri. *US Geol. Surv. Prof. Pap.*, 1042, 48 pp.

Des Plaines

Illinois

Lat/Long: N42° 3′, W87° 52′
Diameter: 8 km
Age: <280 Ma
Condition: Buried

While the Barringer Crater is certainly the best-known US meteorite crater, the Des Plaines Structure has just as certainly been visited by more people than any other. Located under the very urban city of Des Plaines in northern Illinois, it also has the distinction of underlying one of the world's busiest ports, Chicago's O'Hare International Airport, which is close to the southwest edge of the structure.

The Des Plaines Structure was first detected (in a sense) when nineteenth century well-drilling showed a highly inconsistent rate of success in finding sub-surface water in the area. Wells in some regions were completely waterless, including a well drilled for the city in 1894 to a depth of 600 m. This suggestion of underlying disturbance was eventually followed up with study of over 100 cores taken from drilled water wells.

There are no surface features of the Des Plaines Structure (in fact, there are no bedrock outcroppings in the area, which is characterized by a 30-m deep layer of glacial drift). The structure, as gleaned from cores and wells, consists of a roughly circular area of structurally-disturbed rock with a central uplift, displaced vertically by about 300 m. Surrounding the central core is a depressed ring in which Mississippian and Pennsylvanian rocks are found. Complex faulting is present and quartz grains from the structure show evidence of shock-produced features, including percussion fractures and strain lamellae. A few shatter cones have been found in beds of brittle dolomite.

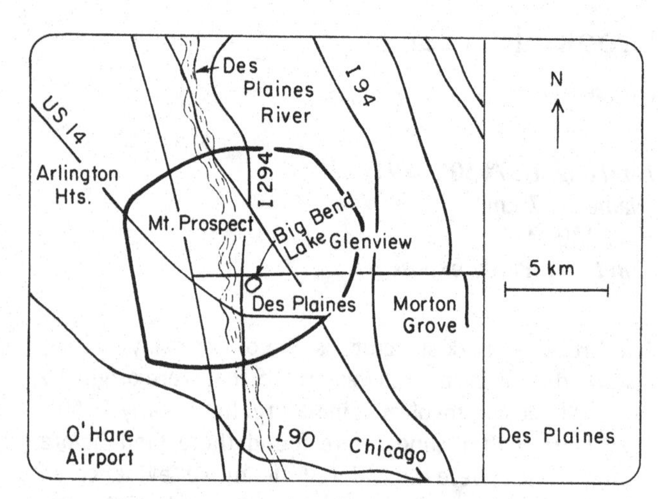

DES PLAINES: Location map for the Des Plaines structure.

Access: Des Plaines can be reached by air from nearly every city in the US, as it almost underlies Chicago's O'Hare International Airport. The center of the structure is under Big Bend Lake on the Des Plaines River. US 14 crosses over it.

References

Emrich, G. H. and Bergstrom, R. E., 1962, Des Plaines disturbance, Northeast Illinois. *Geol. Soc. Am. Bull.*, *73*, 959–968.

McHone, J. F., Sargent, M. L. and Nelson, W. J., 1986, Shatter cones in Illinois: evidence for meteoritic impacts at Glasford and Des Plaines. *Meteoritics*, *21*, 446.

Flynn Creek
Tennessee

Lat/Long: N36° 17′, W85° 40′
Diameter: 3.6 km
Age: 360 Ma
Condition: Partly exposed

The peculiar nature of this feature in northern Tennessee was first noted as long ago as 1869. For most of the first half of the twentieth century it was thought to be a cryptovolcanic structure. However, in 1957 a detailed study suggested that a more likely explanation is that it consists of the highly eroded remnants of an impact crater. The primary characteristic noted at that time was the unusual thickness of a layer of shale (the Chattanooga Black Shale) in an area of otherwise more uniform layering. Intense brecciation is present and shatter cones have been recovered from limestone beds near the center of the formation. The brecciated region occupies an elevated central part of the structure and has been likened to the central peaks of the Sierra Madera Structure and of lunar craters. Extensive drilling has provided a detailed profile of the underlying structure.

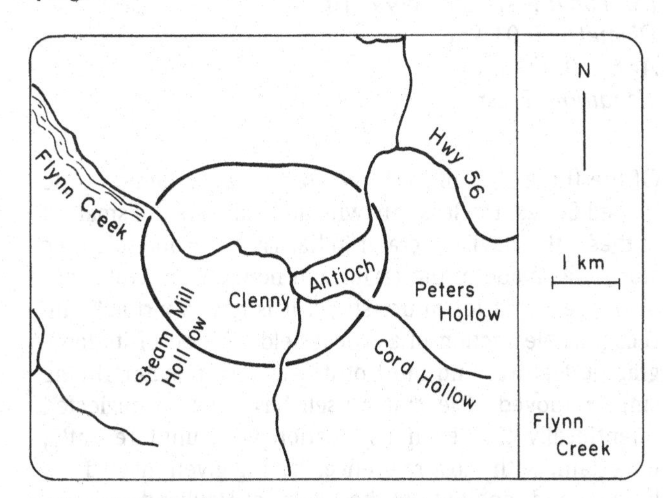

FLYNN CREEK: Location map for the Flynn Creek structure

Access: The Flynn Creek structure is centered on the small town of Clenny, Tennessee, approximately 100 km east of Nashville. It can be reached by following local roads about 20 miles north of Cookeville, north off of Interstate 40.

References

Conrad, S. G., Elmore, R. T. and Maher, S. W., 1957, Stratigraphy of the Chattanooga black shale in the Flynn Creek structure, Jackson County, Tennessee., *J. Tenn. Acad. Sci., 32,* 9–18.

Roddy, D. J., 1979, Structural deformation at the Flynn Creek impact crater, Tennessee., *Lunar Planet. Sci., 10,* 2519–2534.

Glasford
Illinois

Lat/Long: N40° 36′, W89° 47′
Diameter: 4 km
Age: <430 Ma
Condition: Buried

The Glasford structure was recognized as a dome in the 1950s, while exploration for coal was occurring in the region. The top of the dome was found to be at a depth below the surface of 350 m. The local power company took several cores and had a gravity map constructed for the purposes of considering the dome for underground storage of natural gas. These showed that the structure was a local uplift in which denser, lower rocks had been brought closer to the surface. The cores indicated that this uplift consisted of highly deformed rocks, with large, randomly-oriented blocks mixed together in a matrix of finer brecciated material. A complex pattern of faulting further suggested that the dome might be the central uplift of an impact crater. Later discovery of shatter cones in blocks of fractured dolomite places this hypothesis on firmer ground.

Access: The Glasford Structure is located adjacent to the

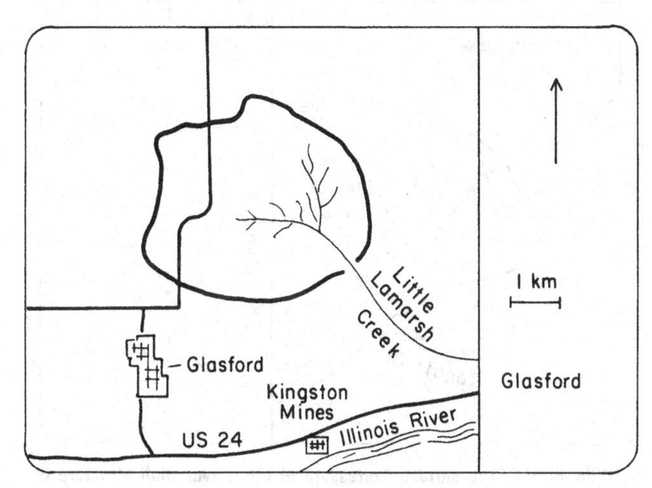

GLASFORD: Location map for the Glasford structure.

northeast boundary of the town of Glasford, Illinois and is 16 km southwest of Peoria. Highway US 24 passes just south of the structure. There is nothing in the way of surface topography to indicate its presence.

References

Buschbach, T. C. and Ryan, R., 1963, Ordovician explosion structure at Glasford, Illinois. *Bull. Am. Assoc. Petrol. Geol., 47,* 2015–2022.

McHone, J. F., Sargent, M. L. and Nelson, W. J., 1986, Shatter cones in Illinois: evidence for meteoritic impacts at Glasford and Des Plaines. *Meteoritics, 21,* 446.

Glover Bluff

Wisconsin

Lat/Long: N43° 58′, W89° 32′
Diameter: 10 km
Age: <500 Ma
Condition: Partly exposed

Glover Bluff consists of three hills that rise about 35 m above the surrounding countryside. Aside from their geological interest, they have been important economically as the site of quarries that have mined the dolomite of Lower Ordovician age. Originally thought to be the result of graben faulting, the impact origin of the structure was only established in 1983, when shatter cones were discovered at the location of a newly-established quarry at the top of the northern hill. The circular form of a crater is inferred from the dip of the dolomite layer and from the detection of bedrock (sandstones) in water wells in the area.

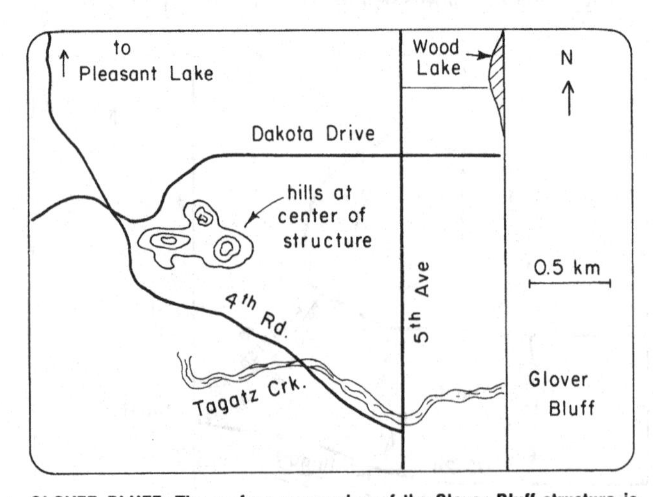

GLOVER BLUFF: The surface expression of the Glover Bluff structure is centered on three hills.

Access: Glover Bluff is located north of Madison, Wisconsin and approximately 20 km north of the Wisconsin Dells. It can be reached by driving 10 km on country roads north from the town of Lawrence or southeast 1.5 km from the south shore of Pleasant Lake.

References

Read, W. F., 1983, Shatter cones at Glover Bluff, Wisconsin, *Meteoritics, 18,* 241–243.

Read, W. F., 1984, The circular structure at Glover Bluff: what and where it is, *Meteoritics, 19,* 295–296.

Haviland

Kansas

Lat/Long: N37° 35′, W99° 10′
Diameter: 0.01 km
Age: <0.001 Ma
Condition: Fresh

Of the three sites of fresh meteorite craters found in the United States, the least known and studied is the smallest of these, the Haviland crater in Kansas. Only one scientific study was made of the crater and none of the crater has been preserved for future study. It is now a virtually unrecognizable section of a wheat field with all of its topographic features and most of its meteoritic material long since removed. The crater itself has been so neglected scientifically that even its location was, until recently, uncertain, with most references having given instead the latitude and longitude of the ‘town’ of Haviland.

Although the Haviland crater itself was recognized relatively recently, evidence of the meteorite’s fall that produced it has been available for more than 100 years. The Brenham meteorites (sometimes referred to in the literature as Kiowa County meteorites) were first discovered in the vicinity in about 1886. Over 1000 kilograms of meteoritic material have been recovered from the area. One fragment, about 500 kilograms in weight, is the largest known and recovered pallasite. Of the thousands of meteoritic fragments, most are highly weathered and oxidized.

HAVILAND: The Haviland crater location is now a plowed field with little to identify it except a few pieces of meteorite shale.

The early discovery of the meteorites was due to the fact that the prairie sod in this part of Kansas is remarkably free of rocks. Therefore, when stones were seen lying on the ground by early settlers, they were conspicuous and remarkable. According to the earliest published report of discovery, the first Brenham meteorites were probably found in 1886 by a cowboy, who came across three large stones, which he called 'strange rocks' and which he buried in a gulch for safe-keeping, unable to take them to the nearest town. He then became ill and died, but on his deathbed divulged the location of the strange rocks to a friend. They were found and taken to Greensburg, Kansas, where apparently at least some realization of their value as meteorites must have occurred. However, one of them, weighing approximately 20 kilograms, was used by one of the settlers to weigh down a haystack and a 50 kilogram one was left in the street in front of a real estate office in Greensburg.

By 1890, many more meteorites had been found, and scientists had been called in to identify them. Meteorites were so plentiful on one farm, owned by the Kimberleys, that it became known in the environs as the 'meteorite farm'. Mrs. Kimberley took an active interest in the rocks found on her farm, collecting them and describing them to visiting scientists. In succeeding years more meteorites were discovered, most of them found in the fields by being hit with plows, mowing machines, corn cultivators and other farm equipment. One meteorite was uncovered by lightning, which was witnessed to strike the soil on a slight rise, kicking up dirt around the meteorite. Another fairly large piece was found in a hole made by some hogs, which had excavated under a barbed wire fence at that point. Mrs. Kimberley herself was using one to hold down her cellar door, and another one to steady the roof of a stable.

Brenham meteorites are still being found, although now in very much smaller numbers. Robert Peck, who was then farming the Kimberley farm, found a 60 kilogram meteorite in 1968 while digging a drainage ditch and a 40 kilogram chunk was found in 1972 about three miles from the crater. A small piece was found by Ellis Peck, owner of the Kimberley farm, at the crater site as recently as 1978. The largest find was made in 1948 by Mr. Peck. It is a 500 kilogram chunk, found west of the crater and now on display in a museum in Greensburg.

The Haviland crater itself was not discovered until about 1925, when H. H. Nininger visited the Kimberley farm to recover the meteorites. From its suggestive appearance, Nininger recognized the crater as being different from the adjoining buffalo wallows. It was deeper, and had a continuous raised rim around it. Furthermore, it held water longer than a buffalo wallow after a rain, and for that reason was a favorite watering place for stock. From Mrs. Kimberley's description, Nininger learned that in the past the rim was much higher than at the time he saw it and the depth of the crater was greater; it had been cultivated for approximately 20 years at that time and much of the topography had been smoothed out. Nininger's measurements showed the crater to be elliptical in shape, 17 m by 10.7 m in size, with its major axis lying west-northwest–east-southeast.

In 1933, under the sponsorship of the Colorado Museum of Natural History, Nininger and J. D. Figgins carried out an excavation of the crater in an attempt to discover its underlying shape and to recover meteorites from it. They found that meteoritic material was located in a cone-shaped layer with a depth of 8 feet at the center and a layer thickness of 12 to 18 inches (approximately 40 centimeters). The deepest meteoritic fragment was found 2.9 meters below the present surface near the center of the crater. The largest meteorite fragment found in the crater was 39 kilograms in weight and in the densest portions the meteoritic material in the meteorite layer gave up as many as 65 fragments per cubic foot of soil. The meteorites were found frequently in small groups, with a large meteorite at the center and small fragments in contact with it, all surrounded by a rust-colored soil approximately 1 centimeter thick.

The Haviland meteorite material is now virtually gone.

It has been completely excavated and almost all meteoritic material larger than microscopic particles has been removed. There is no longer any topographic feature identifiable, as farming activities have now completely filled in the center and smoothed off the rim. It is not possible to look for meteoritic material with a magnetometer, because the farmers, in their attempt to fill in the hole, used it as a dumping ground, burying various things, including scrap metal, for the purpose.

Access: The crater is gone, but the site is easy to reach, though a local guide will probably be necessary to find the exact spot. It is about 15 km east of Greensburg, Kansas, and about two miles south of US 54, several hundred meters from the nearest farm road.

References

Hodge, P. W., 1979, The location of the site of the Haviland Meteorite Crater. *Meteoritics, 14,* 233–234.

Nininger, H. H., and Figgins, J. D., 1933, The excavation of a meteorite crater near Haviland, Kiowa, County, Kansas. *Proc. Colo. Mus. Nat. Hist. 12,* 9–15.

Peck, E., 1979, *Space Rocks and Buffalo Grass.* Peach Enterprises, Warren, MI, 116 pp.

Hico

Texas

Lat/Long: N32° 00′, W98° 02′
Diameter: 9 km
Age: <60 Ma
Condition: Partly exposed

The Hico Structure is a probable impact feature discovered in the 1950s, but not explored geologically until the 1980s. It lies amid nearly horizontal Cretaceous strata of limestones, marls and sandstones. The distinctive lithologies of the several formations make it possible to have good marker beds for mapping the structure.

There is a central uplift about 1 km in diameter, in which beds are exposed that normally lie at least 16 m lower in the surroundings. The center of the central uplift is overlain with layers that have not yet been identified, but extrapolation indicates that the total uplift at the center may be as much as 90 m. A circumscribing set of ring-like depressions, especially conspicuous from aerial and satellite images, surround the central uplift, with the most conspicuous being a ring graben with an outer diameter of 3 km. The outer boundary is a clear ring fault with a downdrop averaging 12 m. An even larger ring-shaped disturbance with a diameter of 9 km is visible on satellite images but not at the surface. The only shock feature reported so far is a shatter – cone-like structure, found by Dan Milton in limestone exposed in a borrow pit in the central uplift.

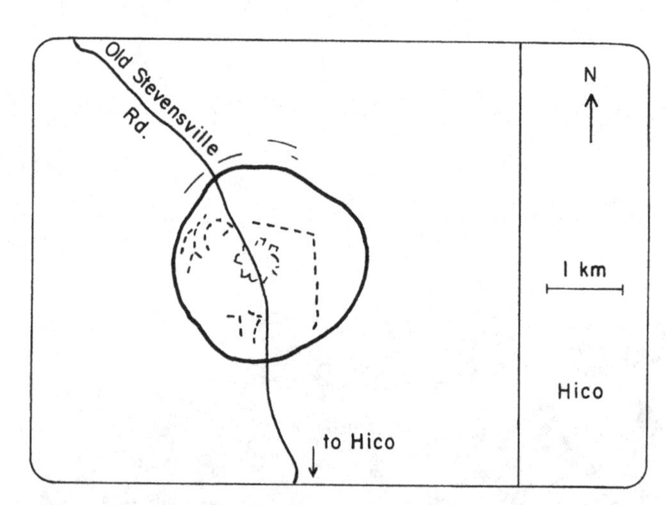

HICO: Sketch map of the Hico structure (after L. W. Milton).

Access: The structure lies in farmland 3 km north of Hico, Texas, 140 km southwest of Dallas. Old Stephensville Road passes through the middle of the structure.

References

Milton, L. W., 1987, in *Research in Terrestrial Impact Structures,* ed. J. Pohl, Vieweg, Braunschweig. pp 131–140.

Wiberg, L., 1980, ‘Bout that big hole north of town. *Lunar Planet. Inst. Bull., 27,* 12–14.

Kentland

Indiana

Lat/Long: N40° 45′, W87° 24′
Diameter: 13 km
Age: <300 Ma
Condition: Partly exposed

Although technically speaking, the Kentland Structure is partly exposed, the amount of rock actually at the surface is very small, almost all of it exposed only because of

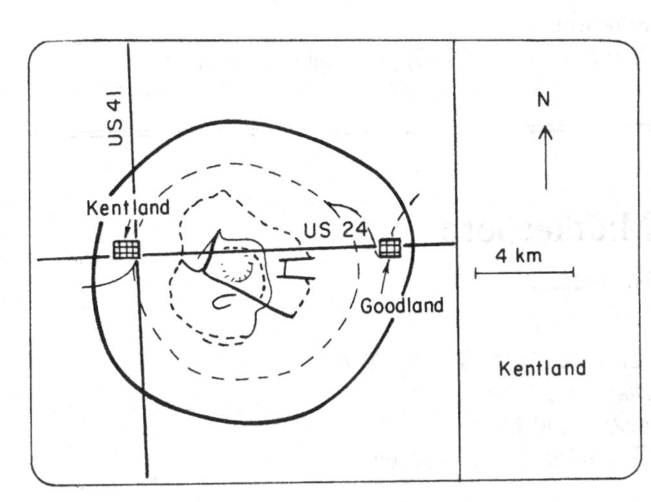

KENTLAND: Location map of the Kentland structure.

quarrying activity. Rocks in the area are covered with about 40 m of glacial till. The bedrock is mostly flat-lying, making the disturbed Kentland structure an obvious anomaly. Underlying rocks are mostly Lower Mississippian and Upper Devonian age sandstone, dolomite, shale and limestone. Knowledge of the sub-surface structure of the feature comes primarily from the exposed walls of the quarry, from geophysical studies, and from drill cores.

The area of disturbed rocks is circular, with a radius of 6.2 km. Inside this radius is a ring-like depression, which surrounds a central uplift. The strata in the central uplift dip steeply away from the axis of the structure. Excellent shatter cones are found in all rocks of the central uplift, and other evidence of shock effects include brecciation and shocked quartz grains.

Access: The Kentland Structure is located in northwest Indiana. US 24, west from Fort Wayne, passes through it. The town sites of Kentland and Goodland lie on the west and the east outer rims of the structure, respectively. The quarry is located near the center of the central uplift.

References

Laney, R. T. and van Schmus, W. R., 1978, A structural study of the Kentland, Indiana, impact site. *Lunar Planet. Sci.*, *9*, 2609–2632.

Manson

Iowa

Lat/Long: N42° 35′, W94° 31′
Diameter: 35 km
Age: 74 Ma
Condition: Buried

In addition to being the largest impact structure in the US, the Manson crater is also one of the best-studied. The reason for this is that its age seemed to place its creation at a time that matched almost exactly with the time of the Cretaceous – Tertiary (K–T) extinction event, which is thought to have been caused by a meteoritic or cometary impact. A better candidate for the K–T boundary impact is the Chicxulub structure in the Yucatan (see Chapter 4). A revised age for Manson, measured in 1993, makes it too old to be involved.

Although the Manson Structure is entirely covered by recent sediments, a wide variety of techniques for exploring it has been employed and the result is a fairly complete picture of its nature. Well drilling provided the indication that there was something unusual under the town of Manson. As early as 1912, its water, for example, was found to be unusually soft for Iowa. Crystalline rocks detected when the Manson city well was drilled were reported to be unusually close to the surface (at depths of about 380 m), which is now known to be because they are part of the central uplift of the crater. Elsewhere in the region wells showed unusual depths for the various sub-surface strata, and it was gradually realized that a large disturbed area was present. Drill cores obtained in the 1940s and 1950s confirmed the remarkable nature of the feature, showing highly disturbed sequences of strata. The pres-

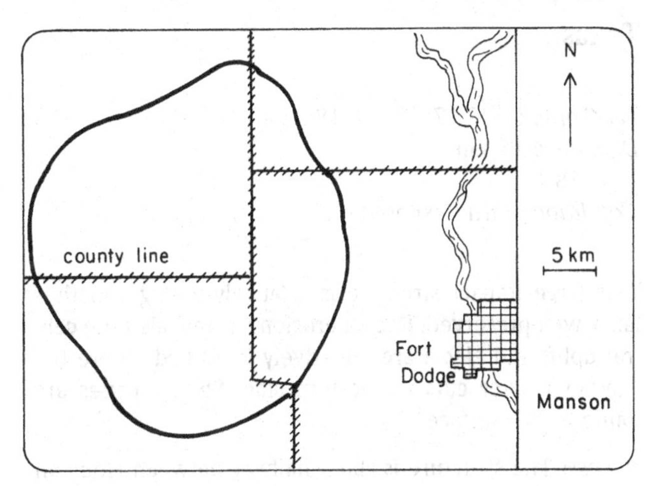

MANSON: Location map of the Manson structure.

ence of shocked quartz from the cores was the first clear evidence that it might be an impact structure.

Gravity, magnetic and seismic surveys all confirmed the character of the structure as a shallow, circular, basin-shaped area of deformed rocks. An uplifted outer rim surrounds a ring of disturbed strata. Interior to this is a ring of highly-disrupted strata, which, in turn, surrounds a central core of igneous and metamorphic rocks. The core includes rocks that must have been brought up from depths of about 6000 m below the pre-impact surface, according to their present depth outside the crater area.

The age of the Manson Structure, which is crucial to the arguments that claimed a connection with the K–T boundary, has been measured by Argon isotope analysis.

Access: The Manson Structure is in north central Iowa, beneath the town of Manson and just west of the city of Fort Dodge. There is nothing to see at the surface.

References

Hartung, J. B. and Anderson, R. R., 1988, A compilation of information and data on the Manson impact structure. *LPI Tech. Rpt. 88–08*, Lunar and Planetary Institute, Houston, 32 pp.

Izett, G. A., Cobban, W. A., Obradovich, J. D. and Kunk, M. J., 1993, The Manson impact structure: $^{40}Ar/^{39}Ar$ age and its distal impact ejecta in the Pierre Shale in Southeastern South Dakota. *Science, 262*, 729–732.

Marquez Dome

Texas

Lat/Long: N31° 17′, W96° 18′
Diameter: 15 km
Age: 58 Ma
Condition: Partly exposed

This large impact structure is relatively young and thus fairly well-preserved. Recent erosion has revealed the central uplift and ejecta are tentatively identified where the Navasota River cuts its western side. Shatter cones are found at the surface.

Access: The structure is about halfway between Houston and Dallas, near Marquez, Texas, which is 25 km on State 7 west of Interstate 45.

References

Sharpton, V. L., 1990, personal communication.

Middlesboro

Kentucky

Lat/Long: N36° 37′, W83° 44′
Diameter: 6 km
Age: <300 Ma
Condition: Partly exposed

Unlike many of the impact structures in the Midwest, the Middlesboro Structure is particularly conspicuous, even though it is both old and highly eroded. It stands out because it is an anomalously flat area amidst a surrounding of mountains and steep stream valleys. Adjacent to the famous Cumberland Gap National Historical Park and housing the entire city of Middlesboro (population ~ 12,000; officially, the correct spelling is supposed to be Middlesborough, though it is seldom used), the structure interrupts the sequence of hills and mountains of the Appalachian chain.

The rocks of the structure and its surroundings are primarily sandstones, quartz-pebble conglomerates, shales and coal, all of Pennsylvanian age. Mining was the principle economic activity of the city, which was originally an iron and steel center and more recently a center of coal mining activity. The impact structure presently consists of encircling faults, a gentle synclinal zone and a central uplift. About half of the structure has outcroppings and half is covered with alluvium from several streams that cross the structure. The rocks of the central uplift are

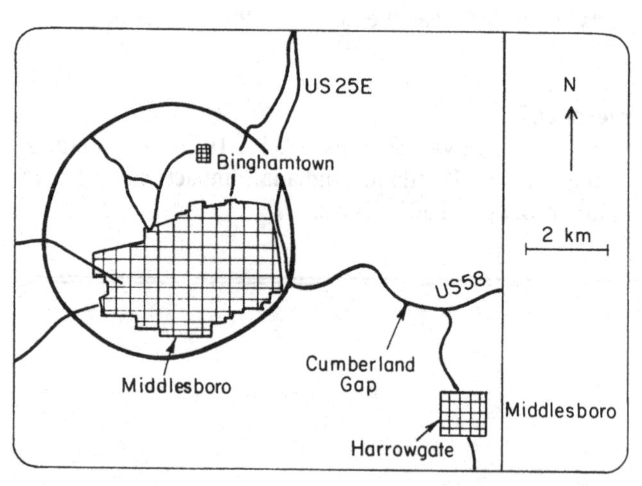

MIDDLESBORO: Location of the outline of the Middlesboro structure.

older than those of the surrounding portions of the basin and have been raised about 300 m above their normal level. The largest outcrop occurs at the corner of the clubhouse of a golf course. It consists of vertically-dipping sandstone that has striations that are similar to shatter-coning. Good shatter cones are found a few hundred meters away near the boundary of the central uplift, where intensely shatter-coned siltstone has been discovered. Additional evidence of the impact origin is the presence of shock-induced cleavage and shattering in quartz grains in the sandstone.

Access: The Middlesboro Structure is easy to visit; it lies beneath the city of Middlesboro, which is near the place where Kentucky, Tennessee and Virginia come together.

References

Dietz, R. S., 1966, Shatter cones at the Middlesboro structure, Kentucky. *Meteoritics, 3*, 27–29.

Englund, K. J. and Roen, J. B, 1962, Origin of the Middlesboro Basin, Kentucky, *US Geol. Surv. Prof. Pap.*, 450E, 20–22.

Odessa

Texas

Lat/Long: N31° 45′, W102° 29′
Diameter: 0.17 km
Age: <0.05 Ma
Condition: Fresh, largely excavated

The largest of the Odessa craters was found by a local rancher in 1892, but their true nature was not recognized until 1922. Large numbers of meteorites were recovered from the rim of the crater, where they were concentrated. The largest weighed approximately 150 kg. They were easy to locate, being iron meteorites (coarse octahedrites). The smallest meteorites recovered are microscopic spherules, a few tens of microns in diameter, that were recovered from samples of the soil in 1965.

The largest crater has a slightly raised rim, rising about one meter above the surrounding flat plain. The center of the crater is about 6 m below the rim. Extensive excavation activity began in 1939, leading to a geologic map of the crater and the discovery of three additional, small craters. The largest of these secondary craters lies 25 m from the primary crater and is about 25 m in diameter. It was found to have a similar structure to the main crater and to have meteoritic fragments primarily limited to its rim and outer flanks. Three additional craters, each about 3 m across, were found about 70 m away. One of these had several large meteoritic masses in its center, about 2 meters below the surface.

ODESSA: The south side of the main Odessa crater from the south rim.

ODESSA: The north rim of the main Odessa crater, showing tilted strata.

ODESSA: Crater 2 and trenches from excavations (foreground).

The excavations made at the main crater included a shaft 50 m deep, which was made in order to search for the expected giant meteorite (at the time it was not yet realized that an impact of this magnitude would virtually destroy the incoming body by its explosive force). No large meteorite was found within the crater.

Because of the exhaustive excavations and owing to periods of neglect (it is said that for a while the county used the crater as a dump!), the site is in rather sorry shape. Fortunately, a local group of enthusiasts who realized its scientific importance bought the surface rights to the area of the main crater in 1963. It was fenced off and a small public museum (now closed) was built. Oil is abundant in the neighborhood and active wells crowd the perimeter of the site.

Access: The Odessa Craters are located 8 km southwest of the city of Odessa, 3 km south off US 80.

References

Barringer, B., 1967, Historical notes on the Odessa Meteorite Crater. *Meteoritics*, *3*, 161–168.

Smith, T. R. and Hodge, P. W., 1993, Microscopic meteoritic material surrounding meteorite craters. *Meteoritics*, *28*, 439.

Sellards, E. H., 1940, Odessa meteorite crater, *Geol. Soc. Am. Bull.*, *51*, 1944.

Red Wing

North Dakota

Lat/Long: N47° 36′, W103° 33′
Diameter: 9 km
Age: 200 Ma
Condition: Buried

The Red Wing Structure (also known as the Red Wing Creek Structure) is completely buried under younger materials. The central uplift has commercially producing oil wells. Cores, seismic evidence and the presence of shock metamorphism all point to an impact origin.

Access: The Red Wing Structure is located in western North Dakota near the north unit of Theodore Roosevelt National Park and just north of the Little Missouri River. It is about 15 km northeast by country road from Stevenson School.

References

Sawatzky, H. B., 1977, Buried impact craters in the Williston Basin and adjacent areas, in *Impact and Explosion Cratering: Planetary and Terrestrial Implications*, pp. 461–480. Pergamon Press, Oxford.

Serpent Mound

Ohio

Lat/Long: N39° 2′, W83° 24′
Diameter: 6 km
Age: <320 Ma
Condition: Partly exposed

The Serpent Mound Structure is characterized by a circular fault zone that surrounds disturbed strata and by a central uplift. Its impact origin is indicated by the presence of shatter cones and coesite.

SERPENT MOUND: The Serpent Mound structure lies below these American Indian earthworks (courtesy of Ohio Historical Society).

Access: Serpent Mound is located 5 km southwest of Sinking Spring in southern Ohio. It is located beneath a wooded area that includes the Native American earthworks of Serpent Mound State Memorial.

References

Cohen, A. J., Reid, A. M. and Bunch, T. E., 1962, Central uplifts of terrestrial and lunar craters: 1. Kentland and Serpent Mound Structures. *J. Geophys. Res.*, *67*, 1632–1633.

Dietz, R. S., 1960, Meteorite impact suggested by shatter cones in rock. *Science, 131*, 1781–1784.

Sierra Madera

Texas

Lat/Long: N30° 36′, W102° 55′
Diameter: 13 km
Age: <100 Ma
Condition: Partly exposed

The Sierra Madera Formation is one of the most difficult examples of an impact structure to explain to someone unfamiliar with the field. Instead of a depression in the ground, this crater has the form of a mountain range. Amidst a flat, almost featureless landscape, the Sierra Madera structure breaks the horizon with its tree-frosted hills (Sierra Madera means 'wooded mountains'). The circular range reaches to about 250 m above the undisturbed plains and 500 m above the lowest part of its surrounding basin. The summit is 1402 m above sea level. Six-shooter Draw is a stream bed that lies along the western side of the ring depression and that breaches the outer raised rim on the southwest and northeast.

Sierra Madera's unusual geology was recognized in the 1920s and the suggestion was made that it might be a cryptovolcanic dome. However, drill cores as deep as 4000 m showed no igneous rocks beneath it. In the 1960s surface investigations showed the presence of large numbers of shatter cones, indicating an impact origin. The shatter-coned rocks are easily found, for instance, in the dry stream beds in the upper parts of the central hills. Subsequent investigations, including gravity and magnetic surveys, have confirmed its identification as an impact structure.

The Sierra Madera mountains are made up of a circular central area of breccia, ranging from microbreccia to samples with meter-sized fragments. Surrounding this is a ring of raised and overturned beds of dolomite and limestone, which show much more deformation than do the same strata exterior to the feature. There are radial faults in the dolomite outcrops of the ring and fragments of concentric faults in the outer parts of the structure. Quartz grains show evidence of shock deformation.

Access: The Sierra Madera mountains are in Pecos County, about 30 km south of the town of Fort Stockton. They are easily seen to the east from US 385, which passes through the outermost western parts of the structure.

SIERRA MADERA: The central hills of the Sierra Madera structure. Meteoritical Society vehicles are gathered in the valley.

SIERRA MADERA: Finding shatter cones in a stream bed at Sierra Madera. Two of the field's pioneers are shown here: H. H. Nininger at the right and R. S. Dietz at the left.

SIERRA MADERA: The hills that make up the Sierra Madera formation rise above the plains of west Texas in this view from the northeast.

References

Eggleton, R. E. and Shoemaker, E. M., 1961, Breccia at Sierra Madera, Texas. *US Geol. Surv. Prof. Pap., 424-D,* 151–153.

van Lopik, J. R. and Geyer, R. A., 1963, Gravity and magnetic anomalies of the Sierra Madera, Texas, 'dome', *Science, 142,* 45–47.

Wilshire, H. G., Offield, T. W., Howard, K. A. and Cummings, D., 1972, Geology of the Sierra Madera cryptoexplosion structure, Pecos County, Texas, *US Geol. Surv. Prof. Pap., 599-H,* 1–42.

Sythylemenkat Lake

Alaska

Lat/Long: N66° 07′, W151° 23′.
Diameter: 12 km
Age: 0.01 Ma?
Condition: Exposed

Discovered on Landsat images in 1977, the Sythylemenkat Lake Structure remains a tentative candidate for an impact origin. Alaska has many circular features (glacial cirques, volcanic structures, and tectonically active areas), and it is thus difficult to identify possible impact structures. There are no signs of volcanism or marks of glaciation associated with the lake, and radial and concentric fractures have been discerned. Anomalous amounts of nickel were found surrounding the lake, with no evidence for a normal parent body for the nickel. A magnetic low was mapped directly over the lake, suggesting a subsurface fractured area (the bedrock is mostly metamorphic and igneous).

Access: The lake lies in the midst of the Ray Mountains, west of the Yukon–Prudhoe Highway. It cannot be reached by surface means and is said to be approachable only in mid-winter, by plane.

Reference

Cannon, P. J., 1977, Meteorite impact crater discovered in Central Alaska with Landsat imagery, *Science, 196,* 1322–1324.

Upheaval Dome

Utah

Lat/Long: N38° 26′, W109° 54′
Diameter: 5 km
Age: <65 Ma
Condition: Eroded, partly exposed

The Upheaval Dome impact structure is a spectacular feature of Canyonlands National Park. It, of course, has little to do with the other geological features of the Park, which are mostly formed from water erosion of the flat-lying beds of colorful sedimentary rock.

Upheaval Dome was first studied in 1927 and subsequent investigations identified it as either a salt dome or a cryptovolcanic structure. By the 1970s it was realized that many structures classed as cryptovolcanic were actually impact features, and so new investigations of Upheaval Dome were carried out. By 1984 the new data had confirmed its impact origin and had established that the exposed structure is the basement of the initial crater, the top 2 km of which has been eroded away. The original crater was probably about 10 km in diameter. The visible dome consists of a central uplift and a more-extended dimple-shaped arrangement of younger rocks, all of which show faulting and convergent displacement.

UPHEAVAL DOME: Upheaval Dome, shown here from the air, is located in Canyonlands National Park (courtesy of National Park Service).

Access: Upheaval Dome is easily reached on a Park Service road, which reaches its edge, where there is a picnic area. The Crater View Trail leads from the road end to a spectacular view point and a longer trail, the Syncline Loop Trail, can be followed entirely around the structure. It goes from the picnic area at the 1800 m level and descends to Syncline Valley and the beginning of Upheaval Canyon, which runs from the Dome west to the Green River.

Reference

Shoemaker, E. M. and Herkenhoff, K. E., 1984, Upheaval dome impact structure, Utah. *Lunar Planet. Sci.*, *15*, 778–779

Versailles

Kentucky

Lat/Long: N38° 02′, W84°, 42′
Diameter: 1.5 km
Age: <400 Ma
Condition: Partly exposed

The Versailles Structure was first described in 1962. Subsequent studies have generally agreed that it is probably a highly-eroded impact structure, but there is less definite evidence (e.g., shock effects in the rocks) than for most of the other objects listed here. There are very few outcrops, but it has been established that the structure has many of the physical characteristics of impact scars. There is an enveloping series of arcuate faults, inside of which the strata are folded and steeply dipping. The central uplift is highly brecciated. All of the strata are Middle Ordovician limestones.

In addition to the structural evidence, further suggestion that Versailles is an impact structure comes from gravity, magnetic, and seismic refraction studies, which show that the disturbance is shallow, as is the case for well-established impact features.

Access: The Versailles structure is located 5 km northeast of the town of Versailles, Kentucky and 20 km west of Lexington. A road, known as Big Sink Road, passes almost directly through the middle of the structure.

References

Harris, J. B., Jones, D. R. and Street, R. L., 1991, A shallow seismic study of the Versailles cryptoexplosion structure, Central Kentucky. *Meteoritics*, *26*, 47–53.

Seeger, C. R., 1972, Geophysical investigation of the Versailles, Kentucky astrobleme. *Bull. Geol. Soc. Am.*, *83*, 3515–3518.

Wells Creek

Tennessee

Lat/Long: N36° 23′, W87° 40′
Diameter: 14 km
Age: ~200 Ma
Condition: Buried

The Wells Creek structure is one of the largest found so far in the United States. It was recognized as a possible impact structure in the 1930s, but the prevailing opinion remained that it was a cryptovolcanic feature until shatter cones were reported in the 1960s. Based on geological mapping, a gravity survey, a magnetic survey, and the location and attitudes of shatter cones, it was then concluded that the structure must be an impact crater, primarily buried and heavily eroded. The reconstructed point of impact is located by the shatter-cone directions at somewhat below the present surface. A central uplift is present and it nearly reaches the surface. The central part of the structure shows steeply outward dipping thrust faults and there are radial tear faults in the rim area. Grabens are found around the periphery.

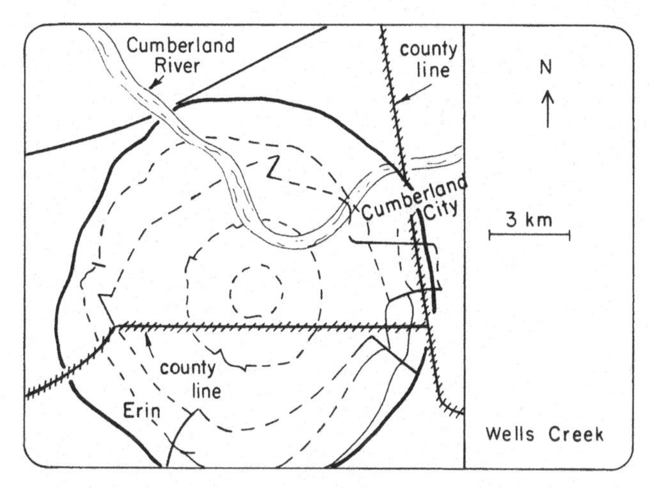

WELLS CREEK: Location map for the Wells Creek structure.

Access: The Wells Creek structure is located about 80 km west of Nashville. The site is overlain by hills of the Tennessee Ridge, Lake Barkley, a bend of the Cumberland River, Big Bend Recreation Area and the Cross Creek

National Wildlife Refuge. The towns of Erin and Cumberland City also lie above it.

References

Stearns, R. G., Wilson, C. W., Tiedemann H. A., Wilcox, J. T. and Marsh, P. S., 1968, The Wells Creek Structure, Tennessee, in *Shock Metamorphism of Natural materials*, ed. French, B. and Short. N. M., pp. 323–338. Mono Book Co., Baltimore.

3

Impact structures of Canada

Brent

Ontario

Lat/Long: N46° 05′, W78° 29′
Diameter: 3.8 km
Age: 450 Ma
Condition: Partly exposed

After the well-publicized discovery of the New Quebec Crater in sub-Arctic Canada, it was realized that the Canadian Shield, that ancient and geologically stable sheet of rock making up most of central and eastern Canada, might retain records of other old impacts. This surmise was confirmed in 1951, when a circular feature was noticed on aerial photographs taken for the Canadian government. This feature, near the settlement of Brent on the Canadian National Railroad, consisted of a depression in which lay two lakes that were separated by a low ridge. The Brent Crater, as it came to be known, is near the northern boundary of Algonquin Provincial Park, a large wilderness park favored by wildlife and canoeists.

A scientific expedition to Brent was immediately successful in discerning its probable impact origin, subsequently well-established by deep drill cores, surface geology, and gravity and magnetic surveys. The lakes, Gilmour Lake and Tecumseh Lake, fill part of the crater basin and the ridge separating them is made up of glacial deposits left by the succession of ice sheets that eroded the rim of the crater, scoured out parts of the interior, and left its debris. The lakes have layers of sedimentary rocks at their bottoms, extending down to 260 m. These

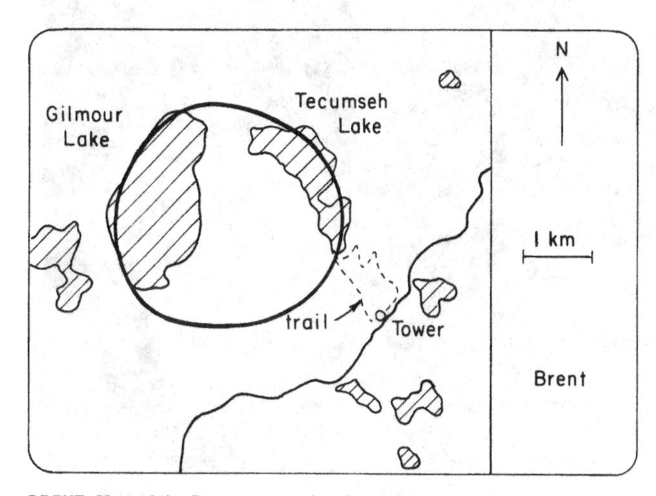

BRENT: Map of the Brent crater, showing the rim's outline and the nature trail to the lake shore.

BRENT: View of the Brent crater from the observation tower, showing the northwest rim of the crater in the distance and Tecumseh Lake at the far left.

BRENT: Tecumseh Lake is slowly being filled in by natural processes, as the lush growth testifies. The original crater depth was about 260 m lower and was filled largely in Permian times.

BRENT: Aerial photograph of the Brent structure (courtesy of Geological Survey of Canada).

rocks, Permian in age, were deposited in the crater when the area was overlain by a shallow sea. Because of the anomalous depth of the crater, these sedimentary rocks have been preserved; they are not found anywhere else in the region.

Drill cores have revealed the deep structure of the crater, which is typical of craters of this size. At the crater center beneath the more recent sediments, lie 810 m of brecciated gneiss, under which is a shallow (40 m) zone of melted rock, formed from the projectile and target material, liquefied at the time of impact. Below the melt zone is a layer of fractured rock extending nearly 1000 m below the surface. The lens of breccia extends to the edges of the crater, thinning in its outer parts. Pieces of this breccia can be found along the remnants of the crater walls and the glaciers have distributed some beyond the crater rim. The writer remembers his first visit to the crater, when after a tiring exploratory tramp through the woods, he sat down on a rock to rest near adjacent Muskwa Lake and discovered that his chair was a nice chunck of brecciated pink gneiss.

Because the Brent Crater is conveniently located inside a well-maintained park, it is being preserved carefully and is being made available for convenient exploration by visitors. There is a wooden tower (built for a field trip of the International Geological Congress in 1972) from which the scale and topography of the crater can be appreciated, and there is a trail from the park road down the inside rim of the crater to Tecumseh Lake. This loop trail gives a visitor an opportunity to glimpse, at least, some of the more interesting features of the structure. The following is a brief guide to the trail (more details can be found in the Strickland reference below):

Distance from tower (km)	Features to note
0.0	The trail leaves the road about 50 m south of the metal marker sign and observation tower. The trail proceeds through the woods down a narrow valley.
0.4	A post (number 2) marks the location of rocks exposed by erosion from deep under the crater rim. They were originally located about 200 m outwards from the inner wall of the crater.
0.6	At the bottom of the trail into the crater is another marker post (number 3). This position, at about the level of the present lake, is near the position of an ancient sea that once filled the crater. The rocks exposed here include an Ordovician talus

Distance from tower (km)	Features to note
	breccia that derived from rocks from the inner crater walls that were washed down into the sea bed, where they mixed with mud and eventually solidified.
0.7	The next marker (number 4) is located near the edge of the present crater floor. The soil and rocks are debris left here by the last glaciation. Beneath it lie layers of Paleozoic limestone with interbedding of sandstone, shale and siltstone. These are not found elsewhere in Algonquin Park, because only here was a depression deep enough to accumulate and preserve these ancient sediments.
0.8	The edge of Tecumseh Lake is reached at marker 5. This and Gilmour Lake are unique in the Park in that both have a nearly neutral pH and are resistant to acid rain. This is because, due to the crater, they are the only lakes in the region with floors of limestone. A spur trail leads out to the water's edge (interesting wildlife, both animal and plant, is abundant here).
1.1	An outcrop at marker 6 shows examples of impact breccia. Most outcrops are so weathered that the vivid pink color of the gneiss is hard to distinguish.

Access: The Brent Crater is easy to visit. A good road leads southwest from Highway 17 at the town of Deux Rivieres. The observation tower and parking area are 32 km from the main highway.

References

Hartung, J., Dence, M. and Adams, J., 1971, Potassium–argon dating of shock-metamorphosed rocks from the Brent Impact Crater, Ontario, Canada. *J. Geophys. Res.*, *76*, 5437–5448.

Millman, P., Liberty, B., Clark, J., Willmore, P. and Innes, M., 1960, The Brent Crater. *Publ. Dominion Observ.*, *24*, No. 1, 43 pp.

Strickland, D., 1987, *Brent Crater Trail.* Friends of Algonquin Park, Whitney, Ontario, 14 pp.

Carswell

Saskatchewan

Lat/Long: N58° 27′, W109° 30′
Diameter: 39 km
Age: 115 Ma
Condition: Partly exposed

The Carswell Structure lies to the south of Lake Athabasca in northwestern Saskatchewan. It was first noticed (in 1956) because of its unusual circular ring of high ridges made up of dolomite. The terrain in the area is otherwise quite flat and uniformly underlain by a layer of sandstone

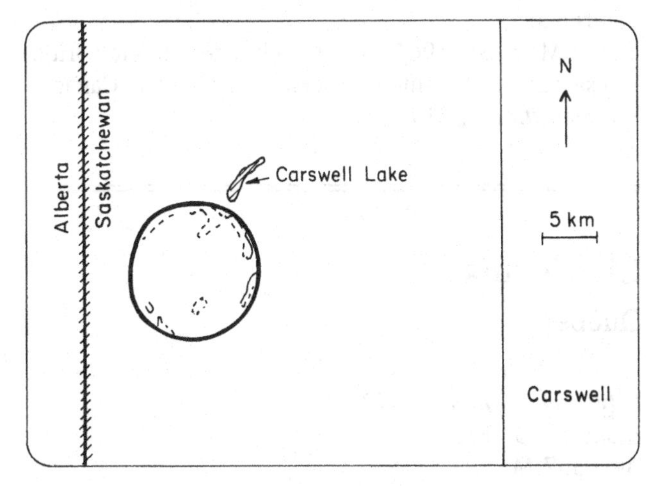

CARSWELL: Sketch map of the Carswell structure (after Innes).

CARSWELL: Satellite photo of the Carswell structure (courtesy of Geological Survey of Canada).

called the Athabasca formation. The dolomite ring, which forms a nearly complete circle, is deformed and fractured, as is some of the adjoining Athabasca sandstone. In the center of the structure the exposures are of gneiss, which represents the central uplift of the impact basin. Shatter cones and other evidence of shock effects have been found in the structure. Magnetic and gravity mapping confirms its probable impact origin. Carswell Lake, which lies across its northeastern edge, is unrelated to the crater morphology.

Access: Carswell is in a remote and barren part of Canada. Maps indicate that the winter-only road from La Loche to Lake Athabasca passes through the middle of it.

Reference

Innes, M. J. S., 1964, Recent advances in meteoritic research at Dominion Observatory, Ottawa, Canada. *Meteoritics, 2*, 219–242.

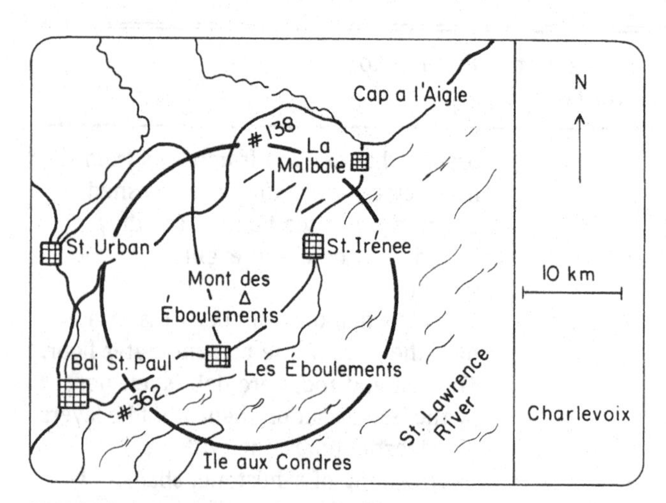

CHARLEVOIX: Sketch map of the Charlevoix structure (after Rondot, 1979).

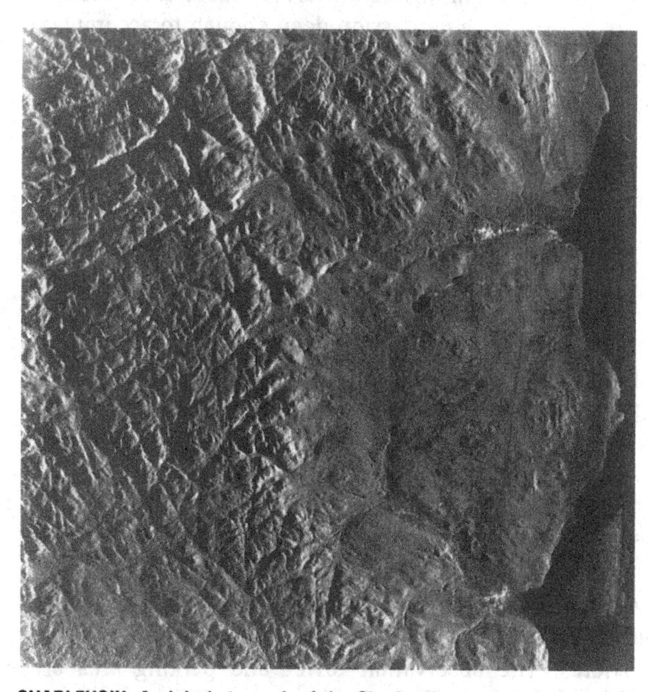

CHARLEVOIX: Aerial photograph of the Charlevoix structure (courtesy of Geological Survey of Canada).

Charlevoix

Quebec

Lat/Long: N47° 32′, W70° 18′
Diameter: 54 km
Age: 357 Ma
Condition: Partly exposed

Quebec has been blessed with an unusually large number of interesting impact structures. One of the most unusual is the Charlevoix structure, which lies half-astride of the St. Lawrence River. It is a very large structure, but apparently the impacting body made a near-direct hit on the giant fault zone that makes the St. Lawrence valley, and only half of the crater's structure survives. It is thought that much of the fused rock formed by the impact, which is now rather scarce at the site, flowed out into the St. Lawrence depression, which existed at the time of the impact.

The form of the Charlevoix structure is typical of impact craters of its size and erosional state, though only half of it is present. There is a central uplift that takes the form of a small mountain, called Mont des Éboulements. Surrounding this is a semi-circular depression that extends out to about twice the radius of the central rise. The original crater has been identified as having its rim at the inner boundary of the outer graben. Evidence for an impact origin is abundant: shatter cones are found at several locations, impact breccia is present, shock metamorphism shows up in quartz, there is mylolisthenite at the inner boundary of the outer graben, and the general morphology is like that of other impact structures of its size.

The Charlevoix structure is one of the few large structures that lie in populated areas where roads provide easy access to its various features. It houses several villages, as well as an island, two small rivers and a spectacular segment of coastline on the St. Lawrence. The following is a suggested tour that can be followed to see a few of the features of the structure's morphology and some of the more interesting outcrops (several good outcrops are on private land and may be visited only with local permission). The tour is based largely on a field guidebook prepared by Jehan Rondot, the structure's primary student (see references below). It begins at the village of Baie St.-Paul, about 100 km northeast of Quebec City on Provincial Highway 138. Just before entering the village from the south, as the highway descends from the Laurentian Uplands, there is a good view ahead of the central peak of the crater, Mont des Éboulements, and ahead and to the left, the valley of the Rivier du Gouffre, which follows the floor of the outer, annular graben. Turn onto Highway 362.

Distance from Baie Saint-Paul (km)	Features to note
2	Recently cut upright beds of Ordovician limestone, which is in contact with sandstone and, farther on, with Precambrian rocks. Turn right on a road that leads to the Cap aux Corbeaux wharf. Walk 50 m east along the railroad track.
4	The cliff bordering the railroad shows an outcropping of a remarkable breccia, called mylolistherite, formed by the movement along the faults that mark the inner boundary of the annular graben. The greatly-fractured gneiss is laced with dikes of mylolistherite. Return to Highway 362 and turn right to Les Éboulements. Pass through Les Éboulements and proceed to a cross-roads. Where the road to Cap-aux-Ories leads right, turn left off Highway 362 and go 1 km to a three-way crossroads.
29	Turn left and drive past an old barn, beyond which there is a gravel pit that has blocks of impact breccia and rocks that show shock metamorphism. From here a private road and trail leads to the summit of the mountain, but permission should be obtained to proceed.
37	Return to Highway 362 and proceed northeast to Saint-Irénee. Park at the railway station and walk northeast along the railroad and then the beach for about 1.5 km. The base of the cliffs here marks the location of the Saint-Laurent fault, while the gullies in the cliffs mark the locations of the intersecting faults caused by the impact. The latter exhibit mylolistherites. The Saint-Laurent rift follows a contact between Precambrian gneisses and the limestones and related deposits of more recent sedimentary rocks that were laid down after the formation of the rift, when erosion of the newly-formed Appalachian mountains brought material from the southwest.
40	Return to Highway 362 and proceed northeast to the crossing of the Gros stream. Outcrops on each side of the stream are of gneiss and amphibolites, with red mylolisthenite dikes present. Glacial deposits lie on an erosion surface, some of which is impact breccia. Continue on Highway 362 through the village of La Malbaie, where you should turn left onto Highway 138. Continue through Clermont and drive about 17 km to the southwest to an intersection with a road to the left leading south-southeast to the village of Saint-Hilarion (1.3 km).
75	Just past the village there is an exposure of rock on the left side of the road. Shatter-coning is prevalent in this rock, which is a charnockite that is cross-cut with pegmatites and thin breccia veins. Return to Highway 138, turn left and proceed to Quebec.

Access: The Charlevoix structure is easily reached from the city of Quebec, following Provincial Highway 138, which parallels the St. Lawrence River to the Charlevoix region.

References

Rondot, J., 1971, Impactite of the Charlevoix Structure, Quebec. *J. Geophys. Res.*, 71, 5414–5423.

Rondot, J., 1979, *Charlevoix Astrobleme, St. Urbain Anorthosite and Stratigraphy.* Department de Geologie, Universite Laval, Ste-Foy, Quebec, 36 pp.

Clearwater Lake, East
Quebec

Lat/Long: N56° 05′, W74° 07′
Diameter: 20 km
Age: 290 Ma
Condition: Partly exposed

Clearwater Lake, West
Quebec

Lat/Long: N56° 13′, W74° 30′
Diameter: 32 km
Age: 290 Ma
Condition: Partly exposed

The Clearwater Lakes occupy contiguous impact basins that have great interest for astronomers, because they were apparently formed by the impact of a double asteroid. Reports of observations of orbiting double asteroids have not always been met with credence, but the Clearwater Lakes indicate that at least a quasi-stable pair must be able to exist in space. The asteroids were not large, probably on the order of 1–3 km in diameter, similar in size to the smaller earth-crossing asteroids being discovered telescopically in large numbers at the present time.

The Clearwater Lakes are unequal in size and not exactly alike in morphology. The larger lake, Clearwater Lake, West, has a circular ring of islands inside it, about halfway out from the center of the lake, while its companion has no such islands and is very deep (about 200 m). Drill cores from the lakes show a similarity, however; both have a basement of fractured and shocked rocks and rock melt, arranged in the characteristic order of impact basins. Gravity surveys confirm their impact origin, as does the fact that shatter cones have been recovered. Both lakes show a central uplift, while Clearwater Lake, West has a clear inner ring, demarcated by the circular array of islands. The Lakes empty to the west into the Gulf of Richmond.

CLEARWATER LAKES: ERTS (an earth resources satellite) orbital photograph of the Clearwater Lakes (courtesy of NASA).

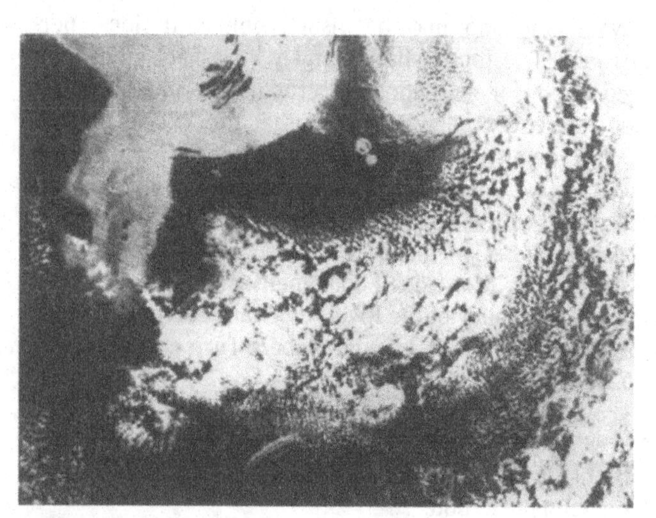

CLEARWATER LAKES: This view of Quebec from space shows the Clearwater Lakes to the right of Hudson Bay and James Bay. Manicouagan is just visible at the bottom of this image, to the right of the center (courtesy of NASA).

Access: The Clearwater Lakes are in northern Quebec, 100 km inland from the eastern shore of Hudson Bay. They are accessible by float plane.

References

Dence, M. R., Innes, M. and Beals, C. S. 1965, On the probable meteorite origin of the Clearwater Lakes, Quebec. *J. R. Astr. Soc. Can.*, *59*, 13.

Deep Bay

Saskatchewan

Lat/Long: N56° 24′, W102° 59′
Diameter: 10 km
Age: 150 Ma
Condition: Eroded

Deep Bay is a remarkable, circular and anomalous feature of Reindeer Lake in Northern Saskatchewan. The main lake is 250 km long, trends northeast (the direction of motion of recent glaciation), is shallow (averaging 30 m) and full of islands. Deep Bay, near its southern end, is, by contrast, circular, deep (averaging 200 m), and devoid of islands.

Although the famous early explorer and fur trader David Thompson had explored the lakes in 1796 and had twice wintered over on Reindeer Lake, neither he nor his successors remarked on Deep Bay's unusual features and it remained unnoticed for the following 150 years. The first public notice of them was made in 1943, when a book describing a canoe trip in the area commented on the oddity of Deep Bay.

The depth of the bay and its lack of islands means that it has unusually treacherous waters during a storm. Apparently this explains the fact that it has held a very bad reputation among the people of the area. There are legends of a monster fish that inhabits the bay, a beast with size and habits not unlike those of the mythical Loch Ness monster. It is said to occasionally break through the ice in the winter to grab a tender, young caribou or two.

In 1951 a Saskatchewanian biologist who had measured the bay's great depth, suggested that it might be a meteorite crater, but the Geological Survey of Canada seems to have dismissed the idea. It wasn't until the discovery of the impact origin of Brent Crater and Holleford Crater that staff of the Dominion Observatory suggested (in 1957) that Deep Bay might have a similar origin. Explorations carried out in the summers and winters of the following few years confirmed this hypothesis.

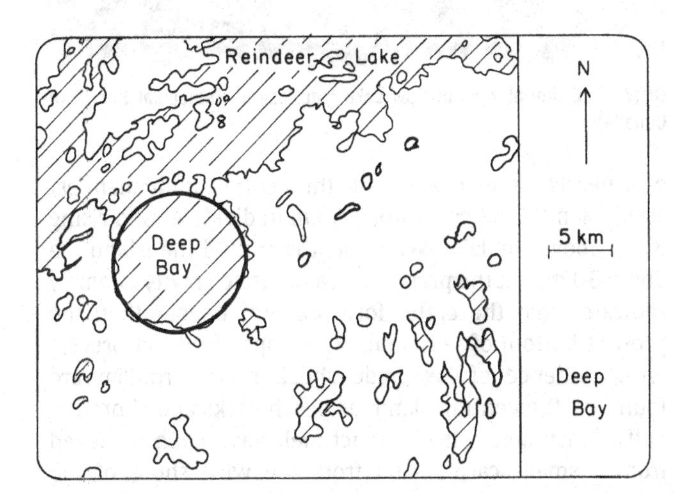

DEEP BAY: Outline map of Deep Bay and Reindeer Lake.

DEEP BAY: Aerial photograph of Deep Bay crater (courtesy of Geological Survey of Canada).

Geological studies showed that the bay is cut out of the Precambrian metamorphic rocks (granite and gneiss) of the Canadian shield and that there exists a layer of unmetamorphosed shale of Mesozoic age in the lake bottom. Although glaciation has eroded the feature conspicuously by grading the rim and scouring the basin, the features of the crater remain rather well-preserved considering its great age. There is an elevated rim some meters high around it and the rim area is cut by numerous radial and concentric fractures, making a conspicuous pattern, both from the air and from any canoe that plies its waters along the steep shoreline.

Drill cores to determine the nature of the lake bottom and the depth of the original crater met with only limited success at first, as it was necessary to do the drilling from the frozen lake surface in winter. There were technical difficulties because of the great depth of the water through which the equipment had to be lowered before it could begin its work in the rock. However, even the early core samples helped to establish the nature of the basin and, when combined with gravity measures, allowed the conclusion to be made that the true depth of the original crater was approximately 1000 meters. The original rim diameter was just under 10 km, according to this reconstruction.

Access: Deep Bay is accessible by float plane from the south or by boat or canoe from trading posts or fishing and hunting resorts on Reindeer Lake. The bay has two broad beaches (where the crater interrupted a broad valley) but otherwise its shores are steep and rocky. Lying just south of the Arctic treeline, the area is nicely wooded with spruce, tamarack, birch and poplar trees.

References
Innes, M. J. S., Pearson, W. J. and Geuer, J. W., 1964, Deep Bay Crater. *Publ. Dominion Observ.*, *31*, No. 2, 52 pp.
Sander, G. W., Overton, A. and Bataille, R. D., 1963, Seismic and magnetic investigation of the Deep Bay Crater. *J. R. Astron. Soc. Can.*, *58*, 16.

Eagle Butte
Alberta

Lat/Long: N49° 42′, W110° 30′
Diameter: 19 km
Age: <65 Ma
Condition: Eroded

Eagle Butte is in the Cypress Hills area of southern Alberta. Cypress Hills Provincial Park occupies part of the eastern portion of the structure. The structure was recognized as a geophysical anomaly long before it was found to be an impact feature. Both surface geology and sub-surface information from wells support its identification as such. The geophysical contours show a central uplift and a surrounding ring depression.

Access: Highway 48 skirts the eastern edge of the structure.

Reference
Sawatzky, H., 1976, Two probable Late Cretaceous astroblemes in Western Canada: Eagle Butte, Alberta and Dumas, Saskatchewan. *Geophysics*, *41*, 1261–1271.

Gow Lake
Saskatchewan

Lat/Long: N56° 27′, W104° 29′
Diameter: 4 km
Age: <250 Ma
Condition: Deeply eroded

Gow Lake is the smallest Canadian crater that has a central uplift. It lies in Precambrian rocks and has the form of a nearly circular lake with the central uplift manifest as an island (Calder Island), 1.5 km in diameter and rising 35 m above the lake. Water depths around the island are about 30 m, but the presence of a negative gravity anomaly indicates that the crater form extends deeper than the present bottom of sediments. Outcrops of impact breccia with evidence of low-grade shock metamorphism are found on the central island, as beach rocks and shoreline cliffs. Small amounts of impact melt have been retrieved from a small scarp inland from the west shore of the island.

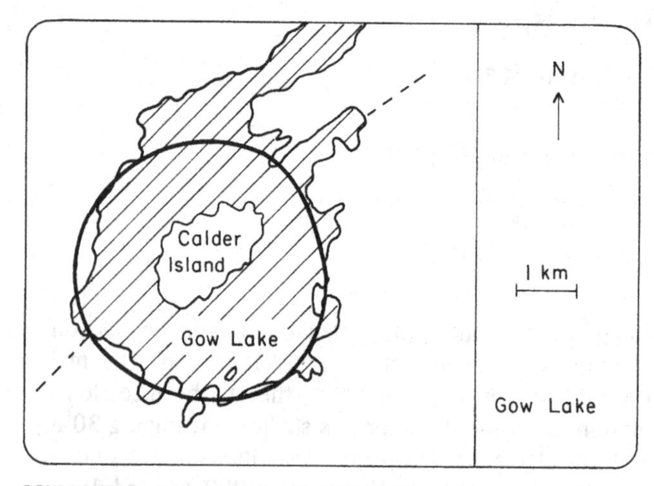

GOW LAKE: Sketch map of Gow Lake (after Thomas and Innes).

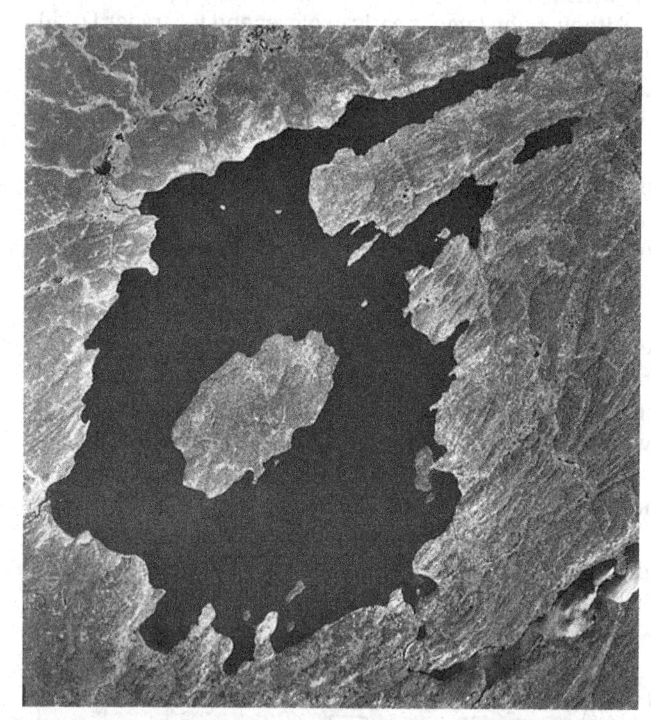
GOW LAKE: Aerial view of Gow Lake (courtesy of Geological Survey of Canada).

Access: Gow Lake is located in roadless, lake-filled country about 80 km west of Deep Bay. A road passes north through this part of Saskatchewan about 50 km east of Gow Lake.

References

Robertson, P. and Grieve, R. A. F., 1975, Impact structures in Canada: their recognition and characteristics. *J. R. Astron. Soc. Can.*, *69*, 1–20.

Thomas, M. and Innes, M., 1977, The Gow Lake impact structure, Northern Saskatchewan. *Can. J. Earth Phys.*, *14*, 1788–1795.

Haughton

Northwest Territories

Lat/Long: N75° 22′, W89° 40′
Diameter: 20.5 km
Age: 21 Ma
Condition: Partly buried

The Haughton impact structure, also known as Haughton Dome, is located on Devon Island, high in the Canadian Arctic. It was first explored in the 1950s but its true nature was not established until 1975, when shatter cones were retrieved from the structure.

The crater lies near the northern shore of the island about 10 km southwest of Thomas Lee Inlet. It is in the western part of the island, where the geological province is the Arctic Platform, made up of marine sedimentary rocks. The eastern end of the island, which is mostly under ice, has exposures of the older Canadian Shield metamorphic rocks. The impact penetrated into these sedimentary beds, exposing the various formations in the impact breccia. It reached depths of at least 1700 m, the depth of the basement metamorphic rocks, as fragments of these are also found in the exposed breccia.

HAUGHTON: From the alluvium-filled slopes on the east side of the Haughton structure, hills of impact breccia are seen to rise about 70 m above the foreground (courtesy of T. Frisch).

HAUGHTON: Shatter cone from the Haughton structure (courtesy of Geological Survey of Canada).

The form of the structure is normal for its size, with a central uplift, a surrounding depressed area and an irregular, highly eroded rim. From the outside, one sees hills of impact breccia rising above broad alluvial valleys. Annular and radial faults characterize the central areas and the rocks at the central uplift lie about 500 m above their normal, exterior position. The faulting is asymmetrical, with the greatest amount in the eastern half of the crater.

In addition to shatter cones, other evidence of shock effects includes the presence of coesite in samples of gneiss and abundant vesicular and flow-structured quartz glass.

Access: Although the island of Devon is not easy to visit casually, it does lie adjacent to Cornwallis Island, where there is an airport that serves the government station at Resolute.

References

Frisch, T. and Thorsteinsson, R., 1977, Haughton astrobleme: a Mid-Cenozoic impact crater, Devon Island, Canadian Arctic Archipelago., *J. Arct. Inst.*, *31*, 108–124.

Robertson, P., 1988, The Haughton impact crater, Devon Island, Canada, *Meteoritics*, *23*, 181–184.

Robertson, P. and Mason, G., 1975, Shatter cones from Haughton Dome, Devon Island, Canada., *Nature*, *255*, 393–394.

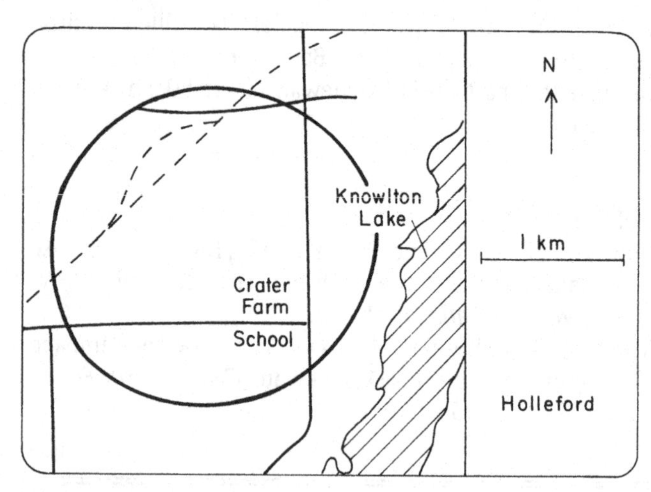

HOLLEFORD: Map of the Holleford area, showing roads mentioned in the road tour.

Holleford

Ontario

Lat/Long: N44° 28′, W76° 38′
Diameter: 2.3 km
Age: 550 Ma
Condition: Partly exposed

On a November night in 1956 a small building on an Ontario farm exploded in flames and burned to the ground. It was, perhaps, a small and inaccurate replay of a spectacular explosion that occurred at this spot 550 million years before. The building had been set up around a drill rig that was exploring down into the strata that had been violently disrupted by the giant meteorite that had created the Holleford meteorite crater. The drill had just reached a depth of 60 meters when it encountered a large amount of natural gas, which raced up the drill hole, encountered the lantern that had been lit for night work, destroyed the building, damaged the equipment, and delayed the Canadian scientist's efforts for a month.

At the time of the explosion the true nature of the Holleford feature was still uncertain. It had been found the year before by two searchers of aerial photographs during a systematic effort to find meteorite craters in the ancient rocks of the Canadian Shield.

The Holleford crater appeared to be a good candidate. From the stereoscopic pair of photographs on which it was discovered, it could be seen to be circular in outline, with a raised rim, and with steeper inner slopes than outer. Vegetation was abundant and roads had been cut through it to the various farms in the area.

When the scientists first visited the feature, however,

HOLLEFORD: The depression that delineates the Holleford crater location, looking north from the barn of the 'Crater Farm'.

HOLLEFORD: A dip in the road near the northern rim of the Holleford depression, looking east.

HOLLEFORD: Shale beds near the northeast rim of the Holleford crater.

HOLLEFORD: The barn located near the center of the Holleford crater.

HOLLEFORD: Aerial photograph of the Holleford structure (courtesy of Geological Survey of Canada).

there was a certain amount of disappointment. Clearly, the stereoscopic view had given an exaggerated idea of the crater-like shape. From the ground, hardly any depression was noticeable at first. The rim was just discernible as a slightly higher area of ground, at most only 25 meters above the low point in the middle. The diameter of the rim was about 2.3 km, but it was not symmetrical, as a drainage valley cut through it, allowing water to flow out to neighboring Knowlton Lake. A grid of farm roads passed through it in a north–south–east–west pattern and the village of Holleford lay partly within it.

Standing at the center of the feature, the scientists were in the middle of a fairly level circular floor about 700 meters in diameter. It was wet almost all year, forming a woodsy bog. To the south rose the best-preserved portion of the rim of the crater, forming a cirque with a distant crest, a kilometer away, rising about 15 m above the floor level. To the east there were shallow scarps, apparently formed by alternating strata that had been eroded because of the protruding rim. To the north were pasture and more of the wooded bog, which snaked its way through a wide gap in the rim to the north and then east to Knowlton Lake. The channel was bordered by cliffs and a mound of glacial deposits lay next to it on the north. If one continued to turn to the left, the northwest side of the rim would be visible; it was also eroded and irregular, but with a scarp that led to a gentle slope towards the center.

An examination of the surface features and rocks of the Holleford area makes it possible to establish the sequence of events that occupied the region's history. Most of the rock in the region is extremely old: Precambrian crystallized limestone and gneiss, both about 600 million years old. On top of this, especially in the southern part of the crater area, lies a layer of younger limestone, dated as Ordovician, about 400 million years old. The crater event must have predated the limestone layer, as the latter fills in the whole central area without noticeable surface disturbance. Wherever outcrops are found inside the crater rim, the strata all slope towards the center, suggesting that the layers were deposited over a circularly-shaped basin.

To determine the true nature of the basin, sub-surface investigations were clearly necessary. Several methods of plumbing below the surface were tried and all gave consistent results, building up a firm picture of what lay below. The first evidence was a magnetic map of the area, which fortunately was already available. It had been produced for this portion of Ontario for the Geological Survey of Canada by flying a magnetometer in an aircraft at an altitude of 150 m, switching back and forth every 0.15 km. The portion of the map that included Holleford did show an unusual behavior for the magnetic field. Elsewhere over the area the magnetic contours varied irregularly and steeply, but within the crater, the pattern was unusually even, with a minimum of contrast. This indicated to the scientists that the crater must be underlain with a very uniform rock layer, which they subsequently identified as the Ordovician limestone layer, deposited in the crater basin during a time when it was covered by the sea.

The next step was to set off a number of explosive charges, allowing the seismologists to estimate depths and thicknesses. The swampy soil passed the sound very slowly, at a velocity of 340 meters per second, but the limestone and the Precambrian rocks both showed velocities of about 5200 meters per second. The times of the echoes were found to be consistent with the presence of a layer of broken-up rock below the present surface, about 100 meters down. This could be impact breccia, which was thought to lie just at and below the original crater floor.

A third way of exploring below ground was then used, involving a network of about 100 gravity stations set up within the crater. These showed that there was a negative gravity anomaly that was symmetrical with the crater rim and that reached a minimum almost exactly at the center of the floor. This could be understood as being caused by the presence of lower-density rock under the crater. The gravity team estimated that the low-density rock formed a layer about 100 meters down, extending as much as 500 meters further. Because the native Precambrian rock and the younger limestone were found to be similar in density, the investigators argued that this must be another type of rock. Impact breccia was a good candidate.

Finally, it was decided to drill down into the crater to bring up specimens of rock that could clinch the case for identifying Holleford as impact crater. Three holes were drilled, ranging in depth from 140 to 450 meters. The cores that were brought up showed that the limestone layers were deposited in a bowl-shaped basin on top of a layer of loose sandstone. Shale was found right at the crater surface and beneath it was a jumble of broken-up rock and breccia.

While the explorers of the Holleford structure were successful in proving its impact origin and in adding to the growing understanding of meteorite craters, others did not fare so well. The farmer on whose land much of the scientific activity took place, Ray Babcock, had been very cooperative and helpful. But the day after the explosion in the night that demolished the drill house, he discovered a remarkable thing in his barn. The well, which had been reliable for as long as he remembered, had suddenly gone dry. Apparently the release of the pressure of the natural gas, which caused the fire and explosion, had changed the balance of pressures underground and no water came now to his 35-meter-deep well. The government scientists attempted to remedy the situation for the unhappy farmer, but the new well they drilled didn't find water until a depth of 70 meters, and then it turned out to be salty, twice as saline as sea water. A third attempt, 200 meters from the original well, was successful at last; fresh water was found at only 25 meters. In their final report the scientists gratefully acknowledged the patience of Mr. Babcock, who surely hadn't known what trouble could arise from trying to farm on land that lay above an ancient meteorite crater.

The Holleford structure is only barely recognizable at the surface. There is little topographic signature and very few outcrops are found. Nevertheless, with some guidance, it is possible to recognize its principal characteristics. The following provides a guided road tour of the surface features of the structure:

Directions	Features to note
Go 3.8 km north of Harrowsmith on road 38	Cross-roads shown on maps as Hartington.
Continue 0.6 km and keep right at intersection	
Continue 0.5 km and turn right at intersection	Pass down a narrow wooded valley to open farmland.
Go west 1.7 km to intersection	The west rim of the crater is directly on the left (north).
Continue west 0.4 km	The slight depression ahead and left leads down to the bottom of the crater. Of course, this depression is not the true crater, but merely a superficial surface expression of the highly-modified original cavity. The bottom of the depression is occupied by a bog.
Continue west 0.6 km	A road cut on the right shows tilted beds of shale.
Continue west 0.7 km	Pass a farmhouse named 'Crater Home'.
Continue west 0.1 km	Pass a barn with a sign labeling it as 'Crater Farm'.
Continue to intersection 0.1 km	A school is on the right and a church on the left.
Turn left and proceed 0.5 km	Directly to the west (left) is the center of the structure, near the northern edge of the low-lying bog.
Continue north 0.5 km	A road cut exposes shale beds on right. The

Directions	Features to note
	depression in the land to the north is the drainage channel that cuts through the northeast part of the rim, allowing drainage into Knowlton Lake.
Continue 0.6 km to intersection, turn left	The intersection lies close to the northeast rim position.
Proceed 0.7 km west	The road has ascended a gentle rise out of the drainage channel and is now on the inner portion of the north rim. To the left (south) is the center of the depression, beyond a wooded area.

Access: The Holleford structure is easily accessible from Kingston, Ontario, which is 28 km southeast of the crater. Drive north from Kingston towards Sharbot Lake and follow the directions in the road log above.

References

Beals, C. S., 1960, A probable meteorite crater of Precambrian age at Holleford, Ontario., *Publ. Dominion Observ., 24,* No. 6, 117–142.

Bunch, T. E. and Cohen, A. J., 1963, Coesite and shocked quartz from Holleford Crater, Ontario, Canada. *Science, 142,* 379–381.

Ile Rouleau

Quebec

Lat/Long: N50° 41′, W73° 53′
Diameter: 4 km
Age: <300 Ma
Condition: Eroded, partially exposed

Ile Rouleau is a peculiar island located near the center of a large, elongated lake in Quebec, Lac Mistassini. It is roughly circular in shape, with a southern extension, and it has a diameter of about 1 km. The shoreline is made up of cliffs about 10 m high on the east and north and a more gentle, rocky shore on the west, while the southern extension is made up of glacial debris. The only outcrops are on the edges of the island, as the center is heavily covered with brush.

Its peculiar geology was first noted in 1973, when shatter cones were discovered at the base of a cliff on its eastern shore. In the following year, a more comprehensive search turned up ample evidence that the island is the central peak of a larger, eroded impact structure.

The principle rock of the island is dolomite, but it is heavily faulted and thin (20 cm) dikes of breccia can be found at several positions around the edge of the island. Quartz grains from the breccia show multiple sets of planar features, indicating a low level of shock deformation. Shatter cones have been recovered from most of the exposures and the apices generally point upward and inward.

In all, these pieces of evidence support the hypothesis that Ile Rouleau is the center of an impact structure that is probably about 4 km in diameter. The island 2 km to the east, Ile Mantounouc, shows no rock deformation, but the arcuate shape of its western shore suggests that the influence of the impact may have extended just about that far.

Access: Ile Rouleau lies near the center of 130 km long lake Mistassini, which is accessible by road from the settlement of Chibougamau.

Reference

Caty, J-L., Chown, E. and Roy, D., 1976, A new astrobleme: Ile Rouleau structure, Lake Mistassini, Quebec. *Canadian J. Earth Sci.*, 13, 824–831.

Lac Couture

Quebec

Lat/Long: N60° 08′, W75° 20′
Diameter: 8 km
Age: 430 Ma
Condition: Partly exposed

Among the thousands of lakes of northern Quebec are a few that are obviously anomalous in shape and origin. The most obvious, of course, is the New Quebec crater, which retains much of its original morphology. Less obvious are the more eroded and altered cases, of which Lac Couture

LAC COUTURE: Mosaic of aerial photographs of Lac Couture (courtesy of Geological Survey of Canada).

is a good example. It is a circular lake among hundreds of rectilinear lakes and a deep, island-free lake among hundreds of shallow, island-studded lakes, and thus is clearly unique in its environment.

The eroded shoreline of Lac Couture is made up of circularly-arranged islets, among which are found examples, especially on the west side of the lake, of impact breccia. These rocks were scoured out of the basin and brought up to the surface level by the intense glacial action that has affected this portion of the Canadian shield so thoroughly. Within the breccia are grains of quartz that show shock-induced micro-fractures.

Lac Couture is 150 m deep at the center and has very little lacustrine sediment overlying the impact breccia at its bottom.

Access: Lac Couture is in a portion of Quebec that is accessible only by float plane.

Reference

Beals, C. S. and Halliday, I., 1965, Impact Craters of the Earth and Moon. *J. R. Astron. Soc. Can.*, *59*, 199–216.

Lac La Moinerie

Quebec

Lat/Long: N57° 26′, W 66° 36′
Diameter: 8 km
Age: 400 Ma
Condition: Deeply eroded

Lac La Moinerie is a lake in northern Quebec with a nearly circular outline. Breccias were found on islands and eskers on the northwest side of the lake. Impact melt fragments have been age-dated. They show various shock metamorphic features.

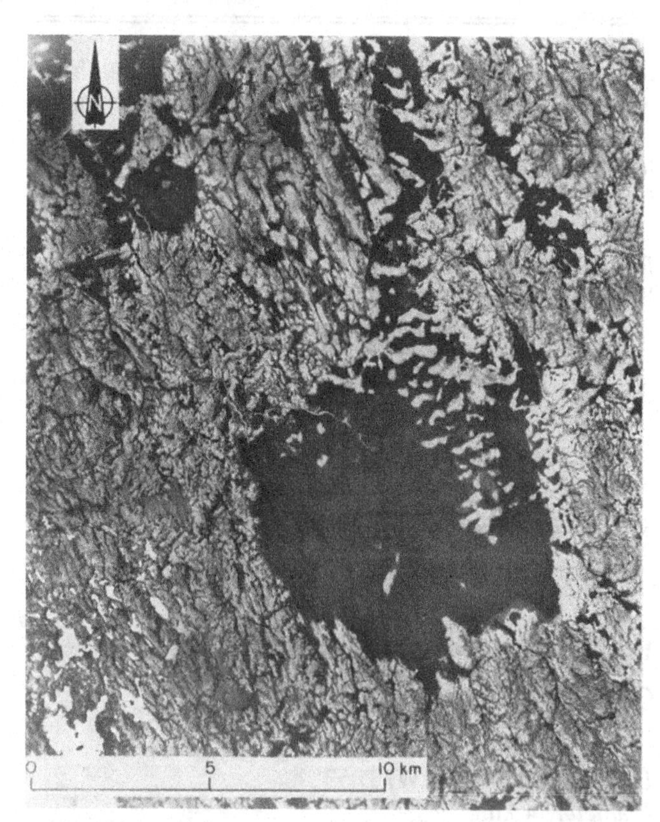

LAC LA MOINERIE: Aerial view of Lac La Moinerie (courtesy of Geological Survey of Canada).

Access: The lake can be reached by float plane.

Reference

Gold, D., Tanner, J. and Halliday, D., 1978, The Lac La

Moinerie crater: a probable impact site in New Quebec. *Geol. Soc. Am.*, Abstracts with Programs, *10*, 44.

Manicouagan

Quebec

Lat/Long: N51° 23′, W68° 42′
Diameter: 100 km
Age: 212 Ma
Condition: Partly exposed

One of the largest impact structures in Canada is a huge circle made by two narrow, semi-circular lakes, Lake Manicouagan and Lake Mushalagan. Together they outline a formation that is conspicuous on even small-scale maps of Canada, and, according to astronauts, is one of the more conspicuous features on Earth as seen from space. The two lakes are connected by a river to the south, which drains them both, eventually emptying into the St. Lawrence near the town of Hauterive.

The center of the Manicouagan structure is a huge central uplift, greatly eroded, of course. It rises 500 m above the surroundings, reaching its peak at Mont de Babel, 952 m above sea level. Immense amounts of shattered and brecciated rock is present and shatter cones have been recovered. The structure includes a 100 m thick layer of impact melt.

Manicouagan is unique in another way: it is presently involved in a large hydroelectric project that has already flooded a large area of the crater and its surroundings. The disadvantage of this circumstance, of course, is that

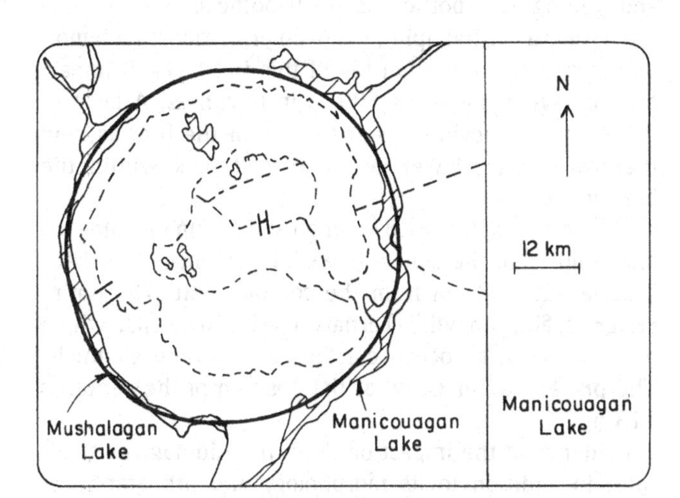

MANICOUAGAN: Sketch map of the Manicouagan structure.

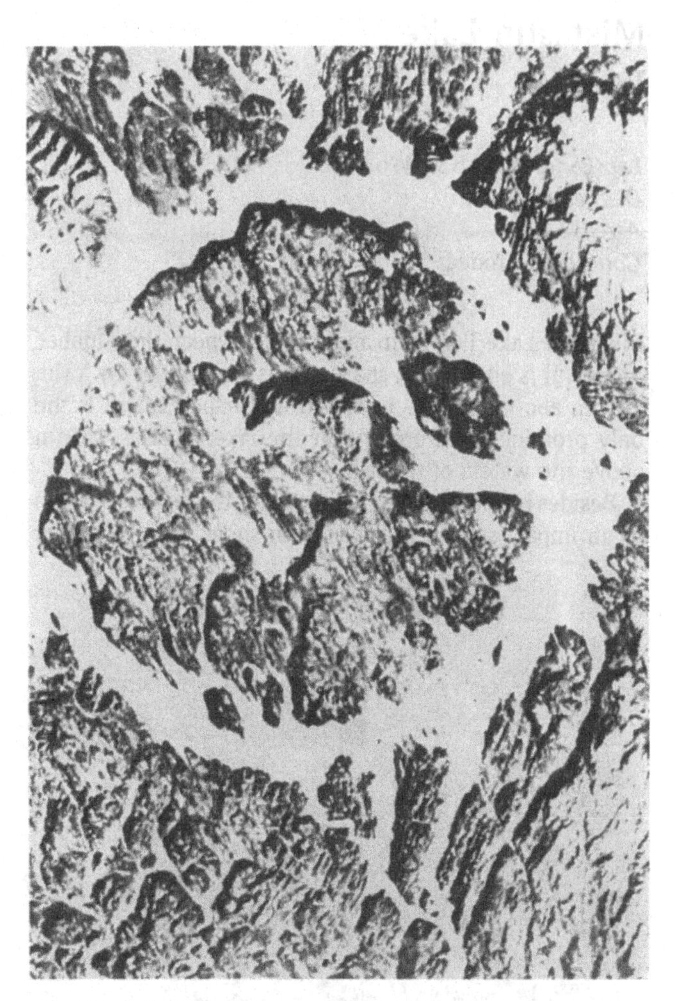

MANICOUAGAN: Orbital photograph of the Manicouagan structure taken in winter, with lakes and valleys frozen and under snow (courtesy of NASA).

some outcroppings of importance become less accessible, while the advantage is that a modern highway leads to the site, potentially providing reasonably easy access to what was until recently a remote and inaccessible place.

Access: Highway 389 leads from Highway 138 at Bae Comeau on the St. Lawrence River nearly 300 km due north to Manicouagan.

References

Dence, M. R. 1964, A comparative structural and petrographic study of probable Canadian meteorite craters. *Meteoritics*, *2*, 249–270.

Mistastin Lake

Labrador

Lat/Long: N55° 53′, W63° 18′
Diameter: 28 km
Age: 38 Ma
Condition: Eroded

Mistastin Lake lies in inland Labrador near the Quebec border. It is elliptical in shape and is surrounded by a rim 150 m above the lake level. An arc-shaped island is the only prominent feature within the lake; it rises 120 m above the waters of the lake.

Besides its morphology, evidence for the feature's origin as an impact structure comes from shock-induced planar features found in the rocks of the central island and the surrounding hills. Igneous rocks that form a prominent hill along the western shore may be examples of impact melt.

MISTASTIN: Satellite view of the Mistastin structure (courtesy of Geological Survey of Canada).

MISTASTIN: Ground-level view of the impact melt sheet at Mistastin (courtesy of Geological Survey of Canada).

Access: The lake may be reached by float plane from settlements in Labrador or Quebec, which are several hundred kilometers distant.

Reference

Taylor, F. C. and Dence, M. R., 1968, A probable meteorite origin for Mistastin Lake, Labrador. *Can. J. Earth Sci.*, *6*, 39–45.

Montagnais

Nova Scotia

Lat/Long: N42° 53′, W64° 13′
Diameter: 45 km
Age: 52 Ma
Condition: Buried

One of the most unusual Canadian impact structures is the Montagnais crater, which lies entirely beneath the surface of the Atlantic Ocean. It was first detected in the mid-1970s, when the continental shelf was being explored for possible oil deposits. When first discovered, the huge circular structure was thought to be an igneous extrusion or an intersection of fault lineaments. However, the possibility of its being an impact crater was also suggested, and subsequent explorations by seismic reflection and drilling have borne out this hypothesis.

There is a central uplift, made up primarily of metamorphic basement rocks. On top of this is a layer of impact breccia, averaging about 500 m in thickness. A layer of impact melt (suevite) lies at the top of the breccia, and there are other, lower layers of melt rock within the breccia zone.

The central uplift is 1.8 km high and 11 km in diameter. Surrounding it, the crater shows the presence of fall back breccia out to 8 km from the central uplift. The entire crater is filled in with Tertiary marine deposits, which have been cut by ocean currents into several channels. The present water depth at the position of the center is 113 m.

Evidence of the impact origin of the Montagnais structure, in addition to its morphology and lithography, is provided by abundant examples of shock-induced min-

MONTAGNAIS: Microphotograph of planar features in quartzite clasts, retrieved from the Montagnais structure (courtesy of Geological Survey of Canada).

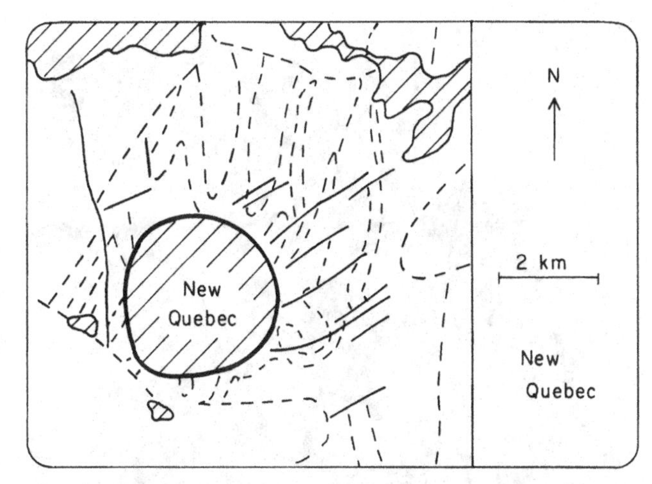

NEW QUEBEC: Sketch map of the New Quebec crater and its surroundings (after Bouchard).

erals (e.g., lechatelierite) and shock lamellae in quartz grains. The impact melt rocks are similar to those found at other large impact sites, such as the Ries. They show an order of magnitude overabundance of iridium, which is probably derived from the incoming stony asteroid or comet.

Access: The Montagnais structure is not easily visited, as it is buried beneath more than 100 m of ocean water and more than 500 m of marine sedimentary rocks. One can pass over it in a boat by proceeding 200 km southwest of Halifax, Nova Scotia, or by going 530 km due east from Hampton, New Hampshire.

References

Jansa, L. F. and Pe-Piper, G., 1987, Identification of an underwater extraterrestrial impact crater, *Nature, 327*, 612–614.

New Quebec

Quebec

Lat/Long: N61° 17′, W73° 40′
Diameter: 3.4 km
Age: 1.4 Ma
Condition: Partly eroded

Previously called the 'Chubb Crater' or the 'Ungava Crater', the New Quebec Crater was the first impact

NEW QUEBEC: Aerial photograph of the New Quebec crater looking north (courtesy of J. D. Boulger, Jr.)

crater to be discovered in Canada. It was reportedly first seen visually by a pilot of a war plane in 1943 and first identified as an impact crater from aerial photographs in that year. It is a simple bowl-shaped crater, only mildly eroded. Paradoxically, the glaciation since formation has done more to preserve it than to erode it.

In structure and shape, the New Quebec Crater is very nearly a scaled-up version of the Barringer Crater. It is, however, filled by a lake, which is 250 m deep near the center. Much of the crater rim is preserved; it rises about 100 m above the surrounding land.

For many years only its shape argued for an impact origin, though by 1962 a gravity map had been made and this further confirmed the hypothesis. Finally, in 1966 a single small rock was found inside the crater rim that turned out to be an example of impact melt (though it was initially thought to be volcanic and evidence against the impact theory). The rock showed evidence in the form

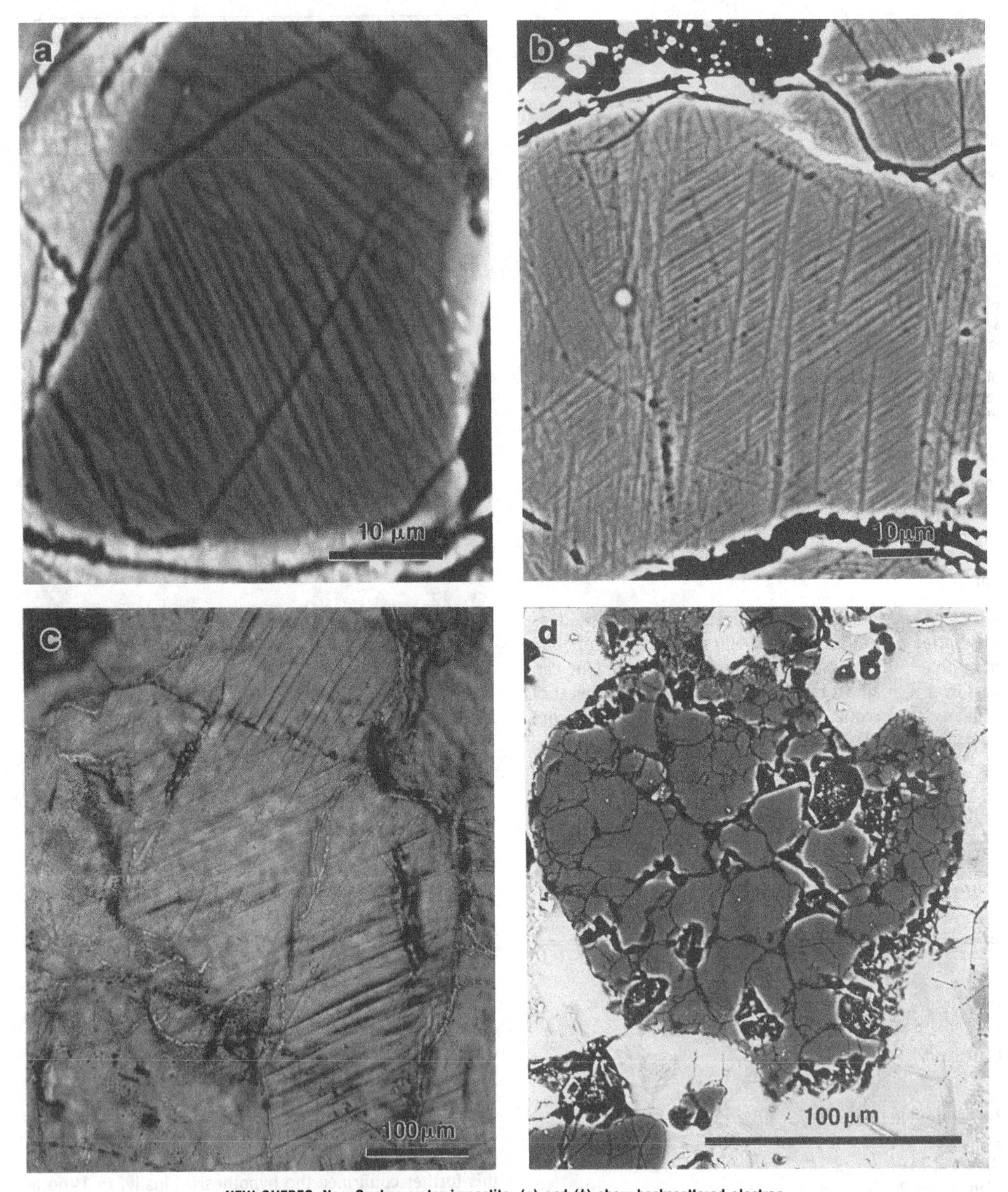

NEW QUEBEC: New Quebec crater impactite: (*a*) and (*b*) show backscattered electron images of quartz grains with multiple sets of planar lamellae; (*c*) shows a photomicrograph of an aggregate of such quartz grains, and (*d*) shows a backscattered electron image of silica with ballen structure, resulting from cooling-caused volume changes (courtesy of Ursula Marvin and the editor of *Meteoritics*, from Marvin, U B. and Kring, D. A., *Meteoritics*, *27*, 585, 1992).

of planar features in quartz and was otherwise very similar to other impact glasses.

Further samples of impact-produced minerals and melt rocks were discovered in 1986 to the north of the crater, where they had been transported by the ice sheet. More were discovered two years later, towards Lac Laflamme, some 3.5 km to the north, the largest example being a melt-rock pebble weighing 650 g. Chemically, the melt rocks are nearly identical to the country rocks; mineralogically they show glass and clasts of quartz and feldspar with shock-induced planar features occurring in some cases. It has been suggested that about 2% of the melt rocks are material from the impacting body, thought to have been a carbonaceous chondrite.

The isolation of the lake and its unusual depth have led to a population of slightly unusual char. Early reports indicated the presence of great monster fish, but systematic studies, involving serious fishing, have demonstrated that the size distribution of the char is normal.

Access: The New Quebec Crater is located in a remote part of the Ungava Peninsula, reachable only by float plane.

References

Bouchard, M. (ed.) 1989, *L'Historie Naturelle du Cratere du Nouveau-Quebec.* University of Montreal.

Grieve, R. A. F., Bottomley, R., Bouchard, M., Robertson, P. B., Orth, C. J. and Attrep, M. 1991, Impact melt rocks from New Quebec Crater, Quebec, Canada. *Meteoritics, 26,* 31–39.

Marvin, U. B. and Kring, D. A., 1992, Authentication controversies and impact petrography of the New Quebec Crater. *Meteoritics, 27,* 585–595.

Meen, V. B. 1951, Chubb Crater, Ungava, Quebec. *Geol. Assoc. Can. Proc., 4,* 49–59.

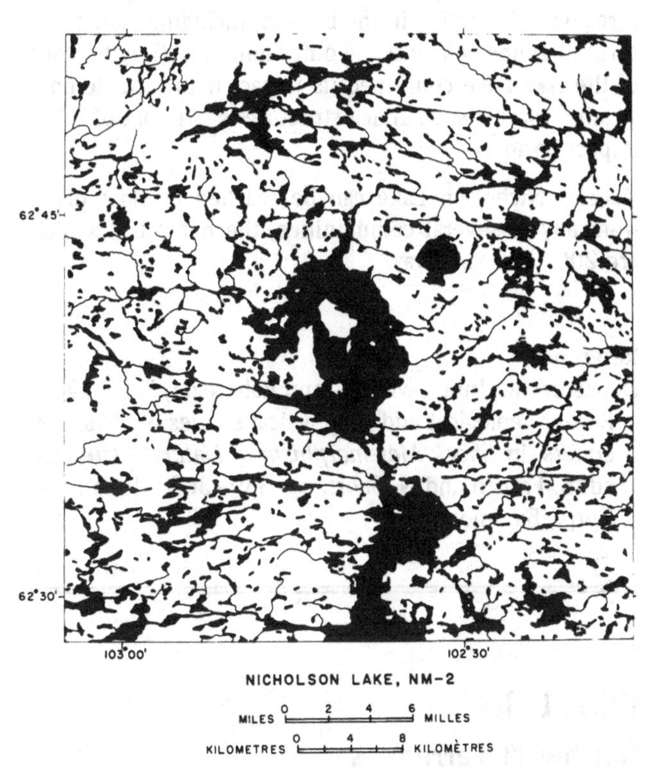

NICHOLSON LAKE: Map of Nicholson Lake, showing the drainage pattern (courtesy of Geological Survey of Canada).

Nicholson Lake

Northwest Territories

Lat/Long: N62° 40′, W102° 41′
Diameter: 12.5 km
Age: <400 Ma
Condition: Deeply eroded

Nicholson Lake, 150 km north of the tree line, is a part of the extensive Dubawnt River drainage system of Northern Canada. The Dubawnt River rises far to the south and flows northward, eventually passing through Markham Lake before connecting onto Nicholson Lake. The river continues northward, eventually emptying into Hudson Bay at Chesterfield Inlet.

Nicholson Lake is only roughly circular and is not an obvious anomaly in its environs, like Deep Bay. It wasn't recognized as an impact structure until 1965, though its peculiar geology had been known for some 70 years. The famous explorer-geologist J. B. Tyrrell had paddled through it and commented on the strange islands found in it. He noted the central island, with its steep outer slopes and the presence of limestone outcrops on another island. The limestone is now known to be Ordovician in age and it is hundreds of kilometers from any similar exposed beds. A further puzzle was the presence of what appeared to be conglomerates and, from 1950s exploration, breccias and tuff, all unrelated to the perfectly normal granites and gneisses of the surroundings.

The mystery of Nicholson Lake was solved in 1965 when evidence for its impact origin was discovered. Shatter cones were found in the gneisses along the western shore of the lake and the rocks previous thought to be conglomerate and tuff were found to be impact breccias of variable character. They are best exposed on a promontory that forms a peninsula at the west side of the lake, and in two central islands. Many examples of shock features were identified in the breccia, including planar features in grains of quartz and other minerals. Gravity maps of the lake have confirmed its impact nature by demonstrating that its sub-surface structure is like that of other impact basins.

Access: Nicholson Lake can be reached by float plane from any of several communities, the nearest of which are 200–300 km away.

Reference

Dence, M. R., Innes, M. J. S., and Robertson, P. B., 1968, Recent geological and geophysical studies of Canadian craters, in *Shock Metamorphism of Natural Materials*, ed. B. French and N. Short, pp. 339–362. Mono Book Corp., Baltimore.

Pilot Lake

Northwest Territories

Lat/Long: N60° 17′, W111° 01′
Diameter: 6 km
Age: 445 Ma
Condition: Highly eroded

Pilot Lake lies in north-central Canada, about 100 km south of Great Slave Lake. It called attention to itself by several unusual characteristics: it is nearly circular but is surrounded by highly elongated lakes; it is free of islands though nearby lakes have many islands; it is not connected to major waterways while other large lakes are; and it is surrounded by a rim of hills 30 to 60 km high, which are unusual in shape and orientation for the area.

Evidence that Pilot Lake is an eroded impact structure was collected in 1965, when it was visited by a scientific mission devoted to checking that possibility. The lake depths were found to be somewhat larger than for neighboring lakes, averaging about 60 m. The bottom was found to be nearly flat and sediment-covered.

The western shoreline of the lake was explored with the hope that glacial erosion might have dredged up rocks from the lake bottom that might indicate impact effects. The rock outcroppings were no help. They appeared to be normal glaciated Precambrian rocks, typical of the region.

However, the scientific explorers found several rocks and boulders near the western lake shore that showed intense alteration, as well as some small pieces of impact breccia. One large boulder of granite gneiss was found lying on a grassy slope, conspicuous because of its highly fractured condition. This boulder was particularly interesting because of its similarity to rocks found around the edge of the Brent crater.

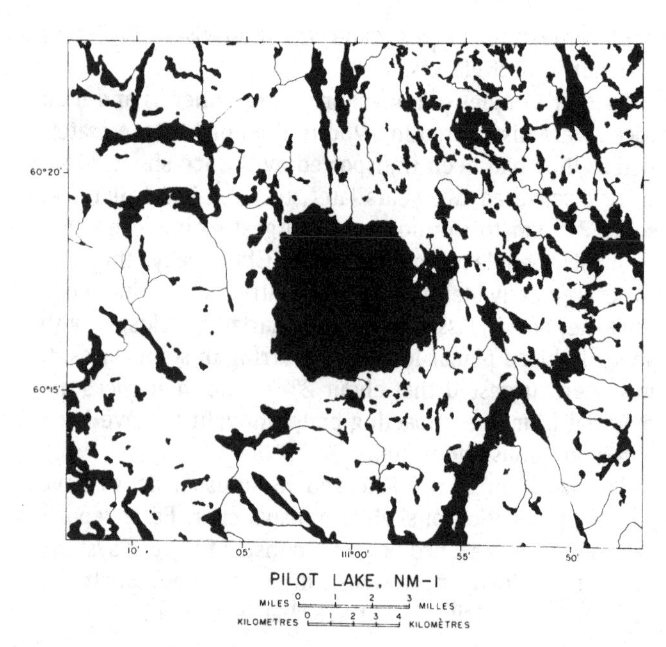

PILOT LAKE: Map of Pilot Lake, showing the drainage pattern (courtesy of Geological Survey of Canada).

Microscopic examination of the rocks showed planar features in quartz crystals indicative of intense shock. Fluidal textures, and recrystalization effects are obvious in the breccias, making them very similar to the Brent breccias, though of more uniform composition. These lines of evidence, together with Pilot Lake's morphology, convinced its investigators that it was formed by meteorite impact.

Access: Pilot Lake is most easily reached by float plane from one of the nearby settlements such as Fort Smith to the southwest or Uranium City to the southeast.

Reference

Grieve, R. A. F., 1991, Terrestrial impact: the record in the rocks. *Meteoritics*, *26*, 175–194.

Presquile

Quebec

Lat/Long: N49° 43′, W74° 48′
Diameter: 12 km
Age: <500 Ma
Condition: Eroded

The Presquile impact structure is centered on a nearly-circular lake known as Lac de la Presqu'ile, which is 7 km in diameter. There is a central peninsula that may be the eroded central uplift of the crater. The main evidence of an impact origin is the presence of shatter cones, which are found on the central peninsula, along the southern lake shore, and about 5 km to the east, along the Obatogamau River. The size of the original crater is highly uncertain.

Access: Aside from the lake, there is little to see of the impact structure at the surface, which is overlain by glacial till. The town of Chapais, Quebec, is a few kilometers to the northwest and the airport at Chibougamau is 35 km to the northeast.

Reference

Higgins, M. and Tait, L., 1990, A possible new impact structure near Lac de la Presqui'le, Quebec, Canada. *Meteoritics, 25*, 235–236.

Saint Martin

Manitoba

Lat/Long: N51° 47′, W98° 32′
Diameter: 40 km
Age: 220 Ma
Condition: Eroded and mostly buried

The Saint Martin structure lies in south-central Manitoba, with Lake Saint Martin lying across its southeast edge. The settlement of Gypsumville is located on top of it. The structure has almost no surface topographic features to suggest its presence. Its nature has been discerned from drilling and from gravity and magnetic surveys, which have revealed a central uplift, a large outer ring of disturbed rock, and a sub-surface layer of impact melt. At the surface there are two outcrops at the position of the central uplift, which shows a vertical displacement of about 200 m. Outcrops of the impact melt are found at three localities around the periphery of the structure. Breccia is found at the outer edge of the ring, but it is covered by post-impact sedimentary rocks. Shock features have been found in the gneiss of the central uplift.

Access: Saint Martin can be reached on Highway 6 from Winnipeg, which is about 230 km to the south. Gypsumville and the structure are about 8 km east of the main highway.

References

Coles, R. and Clark, J., 1982, Lake Saint Martin impact structure, Manitoba, Canada: magnetic anomalies and magnetization. *J Geophys. Res., 87*, 7087–7095.

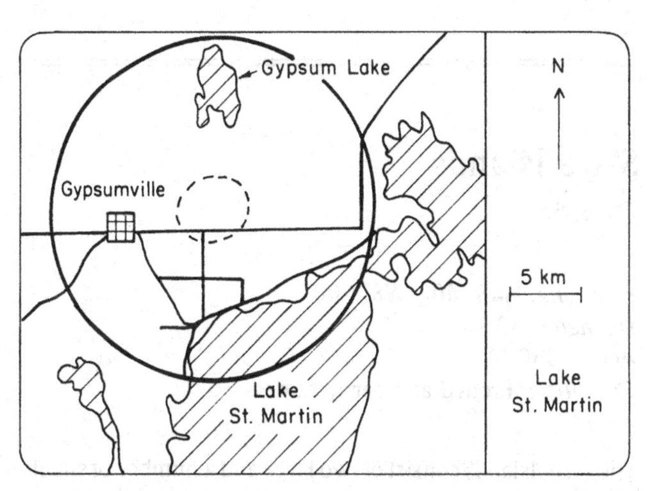

SAINT MARTIN: Sketch map of the Saint Martin structure (after Coles and Clark).

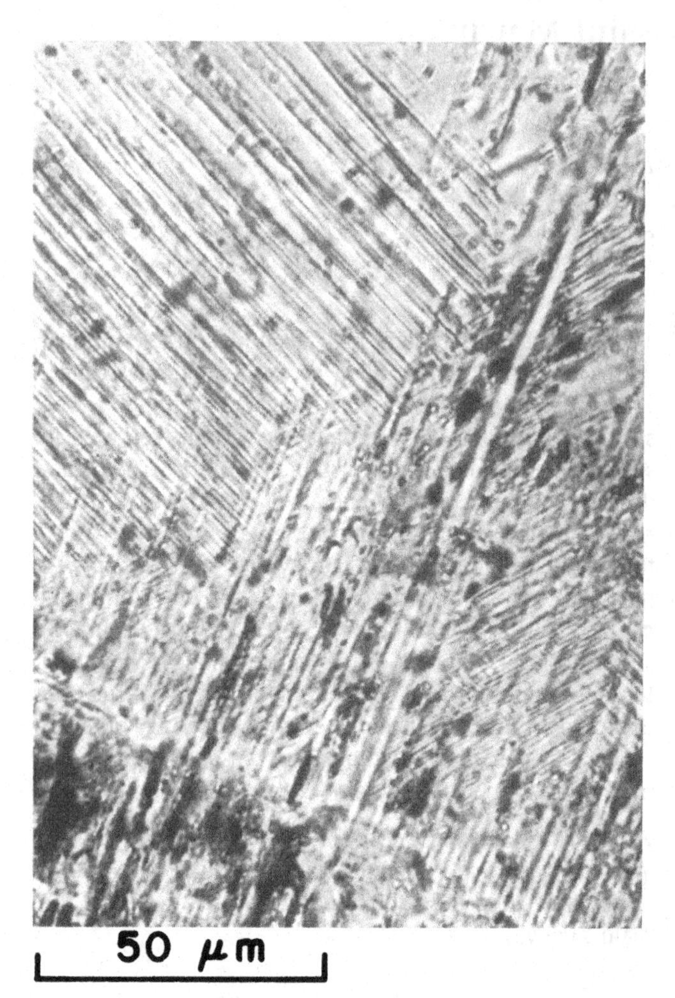

SAINT MARTIN: Planar features in quartz from the Saint Martin structure (courtesy of Geological Survey of Canada).

Robertson, P. and Grieve, R. A. .F., 1975, Impact structures in Canada: their recognition and characteristics. *J. R. Astron. Soc. Can*, *69*, 1–20.

Slate Islands

Ontario

Lat/Long: N48° 40′, W87° 00′
Diameter: 30 km
Age: <350 Ma
Condition: Eroded and partly buried

The Slate Islands consist of two large and a number of small islands lying near the northern shore of Lake Superior. They are about 10 km south of the town of Jackfish, Ontario. Geological and geophysical studies begun in 1973 have shown that the islands are the central uplift of an impact structure, most of which is buried beneath lake waters and sediments. A semi-circular trough with water depths of about 200 m surrounds the islands, except

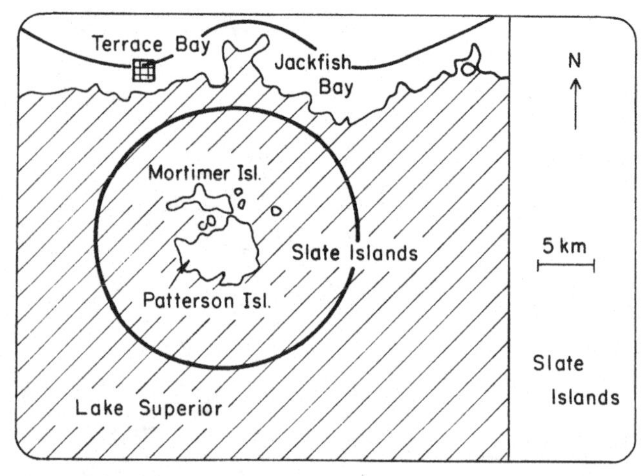

SLATE ISLANDS: Location map of the Slate Islands and nearby mainland.

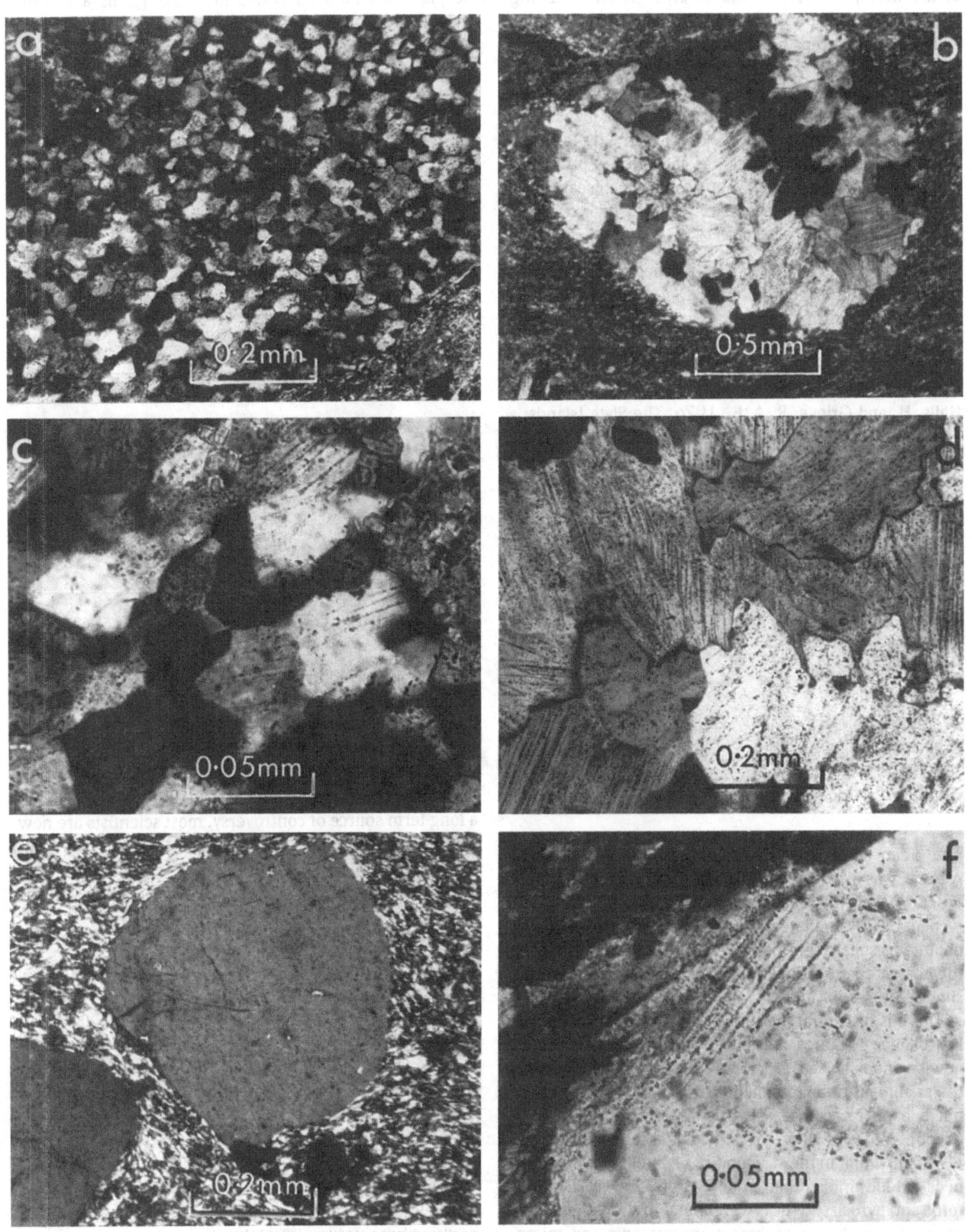

SLATE ISLANDS: Microphotographs of planar deformation features in quartz from the Slate Islands structure (courtesy of Geological Survey of Canada).

at the north, where it is shallower. A submerged ring-shaped ridge lies exterior to much of the trough.

The rock outcrops on the islands show a complicated geological history, including several volcanic events. On the largest island (Patterson Island), the rocks in some places are highly fractured and cut by breccia dikes. Shatter cones are common and a survey of their axes indicates that the center of the impact was above the approximate center of Patterson Island. Microscopic planar features in quartz and plagioclase grains are further evidence of the impact origin of the structure.

Access: The Slate Islands can be reached by private boat from the Ontario mainland (e.g., Terrace Bay) about 10 km to the north.

Reference

Halls, H. and Grieve, R. A. F., 1976, The Slate Islands: a probable complex meteorite impact structure in Lake Superior. *Can. J. Earth Sci., 13,* 1301–1309.

Steen River

Alberta

Lat/Long: N59° 31′, W117° 37′
Diameter: 25 km
Age: 95 Ma
Condition: Buried

The Steen River Structure, near the northern border of Alberta, was discovered in 1963 in the process of a drilling program that was searching for oil. One of the wells turned up a completely unexpected series of rocks at depth and subsequent petrologic study of them disclosed the characteristic structure of an impact crater, as well as shock features in minerals.

There is no hint at the surface that any unusual feature lies below. The area is a huge, flat plain of muskeg. A gravity anomaly had prompted an oil company to drill; the result was disappointment as far as oil was concerned but a boon to science. At a depth of 200 m, the drill encountered crystalline rocks, which extended to depths of 560 m, while in the surroundings flat-lying sedimentary beds extend undisturbed to depths of 1500 m. The plutonic and pyroclastic rocks were found to show the effects of great disturbance on both a macroscopic and microscopic scale. Planar features in quartz grains and other evidence of shock are common.

Access: Although the site can be reached overland from, for example, Great Slave Lake, 150 km to the north, there is no particular reason to visit it, as there is no expression of the structure at the surface.

Reference

Carrigy, M. A., 1968, Evidence of shock metamorphism in rock from the Steen River Structure, Alberta, in *Shock Metamorphism of Natural Materials*, ed. B. M. French and N. M. Short, pp. 367–378. Mono Book Corp., Baltimore.

Sudbury

Ontario

Lat/Long: N46° 36′, W81° 11′
Diameter: 200 km
Age: 1850 Ma
Condition: Partly exposed, distorted

The Sudbury structure is of considerable economic importance in Canada. It contains the world's largest nickel–copper sulfide deposits. Over 15 million tons of nickel and copper have been mined from the Sudbury structure. It also contains a large mining center and good-sized city: Sudbury, Ontario. Although the structure's origin has been a long-term source of controversy, most scientists are now

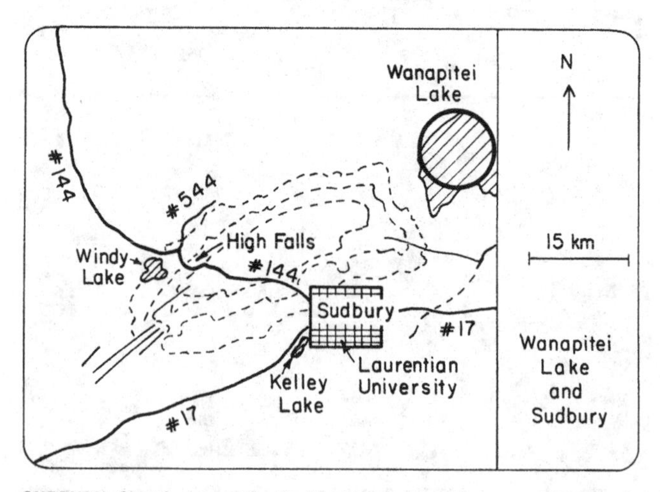

SUDBURY: Sketch map of the Sudbury basin, showing some of the roads mentioned in the road tour (after Lakomy).

SUDBURY: Meteoriticist examining a dyke-like granite inclusion and an olivine diabase dyke (the youngest rock type at Sudbury) in an outcrop of norite.

SUDBURY: Shatter-coning in the Mississagi formation at Sudbury.

convinced that it is an impact scar, into which the ore-bearing rocks, the 'Sudbury igneous complex', was intruded, possibly as a result of the disturbance caused by the impact. The structure, if it once was circular in outline, has been distorted by subsequent geological events into an ellipse with its major axis twice as long as its minor axis.

Both because of the large scale of the impact and because of considerable subsequent rock movement, the Sudbury structure is very complicated. It lies across the boundary of two major subdivisions of the ancient Canadian Shield and thus has been the locus of considerable geological activity. The Superior Province is to its north and the Southern Province to its south. This three-way coincidence of unusual structures has been the leading cause of doubt on the part of some geologists regarding the impact origin of the Sudbury Structure; they argue that its presence on this major boundary is an unlikely circumstance.

Nevertheless, the evidence for an impact is now very impressive. Ring fractures like those around other large impact craters occur north of Sudbury out to about 25 km from the center. Radial faults occur along its inner boundaries. There is evidence south of the crater for an overturned crater rim, where the Huronian formations, which make up most of the rocks south of the crater, are upside – down in some places. Also shatter cones and other evidence of shock metamorphism are plentiful.

Impact breccias are also common. There are two types that have long been recognized by geologists, though their history and origin has only recently been unraveled. One is called the Sudbury Breccia and the other the Footwall Breccia. Surrounding the ring of meltrocks of the Sudbury Igneous Complex are many bodies of Sudbury Breccia, which extend (sparsely) out to distances of more than 30 km from the structure's center. They are most easily found to the south, where rock outcrops are abundant. The breccias are primarily made up of a matrix that is a fine, weakly recrystallized rock flour and fragments that are angular if small, and somewhat rounded if large. The fragments are of the same type of rock as that in their surroundings, with some exceptions. The chemistry of the matrix indicates that there was mixing of material from other parts of the structure and thus that the breccia-forming process involved a large-scale (and violent) event.

The Footwall Breccias are found in relatively thin sheets adjacent to the Sudbury Igneous Complex and generally interior to the Sudbury Breccias. They have a considerable mixture of rock types and have been greatly altered and contaminated by activity subsequent to their formation by the impact event. Not only do they contain shock-induced minerals from their origin but they also have large amounts of rich nickel and copper sulfide ore, apparently introduced by subsequent igneous activity. Contact with the hot rocks of the Sudbury Igneous Complex led to partial (and rather complicated) metamorphism.

An elliptical ring, 5 km thick, 25 km wide and 60 km long, characterizes the surface expression of the Sudbury Igneous Complex. At its outer borders are the Footwall and Sudbury Breccias and at its inner border it grades into the Onaping Formation. It consists of a number of different types of igneous rock: norites, gabbros, and granophyres. Chemical evidence suggests that this material was extruded from a magma chamber below Sudbury that was re-melted and released to the surface by the impact. In the process it mixed partially with the melt rocks at the surface, producing a volcanic rock of unique chemical and physical character.

The rocks of the center of the Sudbury structure are called the Whitewater Group. They lie interior to the ring-shaped Sudbury Igneous Complex. There are three recognized members of the Whitewater Group: the Chelmsford, the Onwatin and the Onaping Formations, in order of their occurrence outward from the center of the structure. The first two of these are considered to be unrelated to the impact event, but rather are sedimentary rocks laid down subsequent to it. The Onaping Formation, on the other hand, is an important source of information about the circumstances of the structure's origin. It is made up primarily of a complicated mix of rock types in a breccia-like mixture, with many special characteristics. Geologists have recognized four different sub-divisions: the Basal Member, the Gray Member, the Black Member and the Melt Bodies.

The Basal Member is generally found in the outer part of the Onaping formation and is mixed in places with the Sudbury Igneous Complex. It is made up of a mixture of rock fragments that can be identified with adjacent country rock, with sizes ranging from 100 m or so down to microscopic size. Shock metamorphism is common. The matrix is inhomogeneous and appears to be a solidification of finely pulverized rock.

The Gray Member lies above and (at the surface) interior to the Basal Member. It is a mixture of rock fragments, glass, pulverized country rock, and mineral fragments and has the look of what one would have if one stirred a mixture of molasses (unpleasantly gray-colored molasses), flour, wheat germ, chopped nuts, marshmallows and crushed rock over a hot stove.

The Black Member is similar in most respects to the Gray Member, except for the presence of a black carbonaceous matrix. Unlike the Gray Member, there is bedding present, though it is discontinuous. The upper parts are fine-grained. Both Black and Gray Members show shock metamorphism in the form of shock melting, planar features in quartz and shatter cones.

The Melt Bodies in the Onaping Formation are irregular lenses of igneous rocks usually found between the Basal and the Gray Member rocks. They apparently were melted rock from the impact event that was intermixed with the fragments and rock dust that fell to the surface after the impact, forming the breccias of the Onaping Formation.

The sequence of events that led to the present, rather complicated structure of Sudbury can be summarized as follows:

Before the impact	A thin layer of sedimentary and volcanic rocks of the Southern Province overlies the archean basement rocks, which are exposed to the north as part of the Superior Province.
Impact	Shock melting of country rocks, ejection of melt rocks and shattered fragments and rock flour, brecciation in a bowl-shaped zone, shock metamorphism in surrounding rocks, and faulting along concentric rings occur.
Minutes after impact	The floor of crater rebounds. The Basal Member of the Onaping Formation rings the crater. Rock, glass and dust are blown into the stratosphere. Footwall Breccia lies just exterior to the central lake of melt rock.
Hours later	Debris rains to the surface to form the Gray and Black Members of the Onaping Formation. The magma chamber below the crater is released and begins to leak up towards the surface.
Later	Sediments accumulate in the crater to form the Onwatin slate and later the Chelmsford turbidites.
Later	Geological activity along the major fault zones that separate the Superior Province from the Southern Province distorts the structure's shape into an ellipse.

Because the Sudbury Structure is so easily accessible and explored, a tour of some its exposed features is given here. The tour begins at the Laurentian University campus, located in Sudbury, and noted for the fact that shatter-coning can be seen in many of the rocks exposed between the buildings. The tour makes a cut through the minor axis of the structure, from the southeast to the northwest.

Distance from campus (km)	Features to note
0	Shatter cones are exposed in Mississagi Formation metamorphic rocks (quartzite) on the campus. Proceed to Highway 69 and south to a side road that runs along the southeast shore of Kelley Lake.
11	Some of the grandest shatter cones are found in the outcrops 4 km from the northeast end of this oblong lake. The best examples are on the south side of the road facing the lake. Some of the cones are as much as 1 m long. On the average the axes of these cones (as well as of most shatter cones found at Sudbury) point towards the center of the basin.

Distance from campus (km)	Features to note
	Return to the northeast and stop where the road passes between Kelley and Robinson Lakes.
16	The outcrop is of quartzites and siltstones on the south and graywacke on the north, with both exhibiting breaches by inclusions of Sudbury Breccias, the best examples being on the south.
	Now drive on to Highway 17 and proceed north to Highway 144. Turn left and proceed north, crossing over the Sudbury Igneous Complex and the rocks of the Whitewater Group, all of which are better exposed to the north (see below). Pass the turn-off to the town of Chelmsford and proceed 3.5 km to the next stop.
43	At this exposure, the Chelmsford sandstone is seen. It lies nearly horizontally, with a slight dip to the northwest.
	Continue west on Highway 144, passing over the slate of the Onwatin Formation. Stop at High Falls on the Onaping River.
57	To see good outcrops of the Onaping Formation, walk along the railroad tracks about 450 m to a cut (see a detailed map). Here the top of the Sudbury Igneous Complex is found (in the form of micropegmatite) and a trail scramble up the hill to the north will take you to the contact between it and the basal breccias of the Whitewater Group. The gray member of the Onaping is exposed here and irregular inclusions of melt rocks can be found. In a sense, this area is the key to the Sudbury mystery, as it was rocks from here that Bevan French studied in 1967, when he found petrographic evidence for shock effects, first demonstrating that Sudbury is an impact structure.
	Continue on Highway 144 to just south of its junction with Highway 544.
61	Outcrops at this stop are all of the Sudbury Igneous Complex. Nearest the road is found an oxide-rich gabbro, while norite is to the northwest. Walk south up the hill to cross onto micropegmatite.

Distance from campus (km)	Features to note
	Continue on Highway 144 through the town of Onaping and on to the railroad crossing north of Windy Lake, past roadcuts through norite.
68	Past the railroad crossing, the highway crosses over the border between the Sudbury Igneous Complex and the ancient footwall rocks, though the contact is not exposed. The gneisses are disrupted and have large inclusions of breccia, with large blocks of country rock in a dark matrix. Beyond this hill the gneiss is progressively less and less disrupted.
	Return to Sudbury on Highway 144.

Access: Sudbury can be reached easily by road or by air. Major Canadian airlines serve the Sudbury Airport, which is located on glacial till that covers this part of the Sudbury Impact Structure.

References

Dence, M. R. and Guy-Bray, J., 1972, *Some Astroblemes, Craters and Cryptovolcanic Structures in Ontario and Quebec.* International Geological Congress, Ottawa.

Dressler, B. O., Morrison, G. G., Peredery, W. V. and Rao, B. V., 1987, The Sudbury Structure, Ontario, Canada: a review, in *Research in Terrestrial Impact Structures*, ed. J. Pohl, pp. 39–68. Vieweg and Sohn, Braunschweig.

French, B., 1968, Sudbury Structure, Ontario: some petrographic evidence for an origin by meteorite impact, in *Shock Metamorphism of Natural Materials*, ed. B. French and N. Short, pp. 383–412. Mono Book Corp., Baltimore.

Wanapitei Lake

Ontario

Lat/Long: N46° 45′, W80° 45′
Diameter: 8 km
Age: 37Ma
Condition: Partially exposed

Wanapitei Lake lies just to the northeast of the large, older and better-known Sudbury Impact Structure. It is an ice-cream-cone-shaped lake, with the ice-cream part marking the northern half-circle of the original crater and the cone being a southern extension of the lake that was glacially carved. The northern half of the lake is deep (~100 m) and island-free. A circular gravity anomaly (which first demonstrated that it is an impact structure in 1972) is centered in the north-central part of the lake and is similar in shape to those of other impact basins.

WANAPITEI LAKE: View from space of Wanapitei Lake (right) and the Sudbury basin (center and left of center) (courtesy of Geological Survey of Canada).

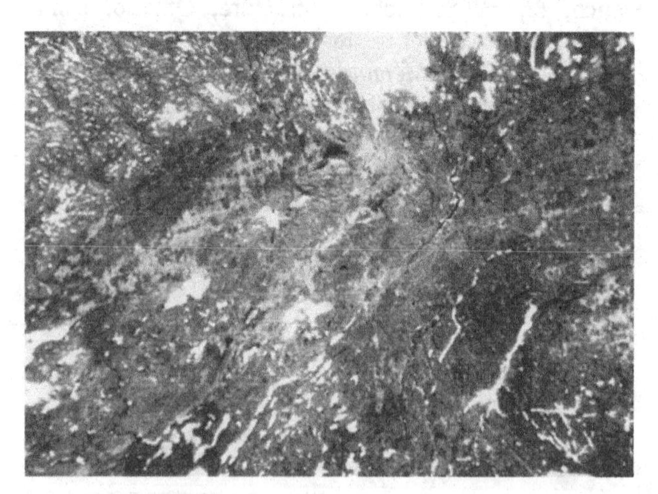

WANAPITEI: Winter view from orbit, showing Wanapitei Lake (top) and Sudbury (upper left) (courtesy of NASA).

Further evidence of its origin can be found in the boulders of soft impact breccia that can be found to the south of the lake. The breccia apparently was scooped up from the bottom of the lake by the glacier and deposited by the southern shores. Shock-induced metamorphism, including vesicular quartzites, coesite, lechatelierite and various examples of fused and glassy rocks are present.

Access: Wanapitei Lake is close to the Sudbury Airport and can easily be visited via a number of good roads. Samples of the soft, friable Wanapitei impact breccia can be found south of the lake, near Bowland Bay.

References

Dence, M. and Guy-Bray, J., 1972, *Some Astroblemes, Craters and Cryptovolcanic Structures in Ontario and Quebec.* International Geological Congress, Ottawa.

Dence, M. and Popalar, J., 1972, *Evidence for an Impact Origin for Lake Wanapitei, Ontario.* Geological Association of Canada, Special Paper No. 10, pp. 117–124.

West Hawk Lake

Manitoba

Lat/long: N49° 46′, W95° 11′
Diameter: 3.2 km
Age: 100 Ma
Condition: Partly exposed

West Hawk Lake is similar in size to Brent, but differs in having water fill the basin entirely. Like Brent, West Hawk

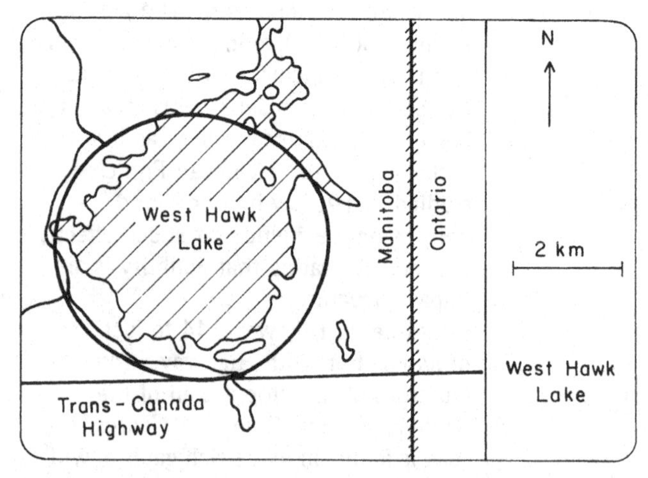

WEST HAWK LAKE: Location map for West Hawk Lake, showing the outline of the impact structure and the location of nearby roads.

WEST HAWK LAKE: Aerial photograph of West Hawk Lake (courtesy of Geological Survey of Canada).

Lake is handy to Canadian population centers and therefore can easily be visited. Its shoreline is highly eroded and irregular, making many bays and inlets. Low surrounding hills may be the skeletons of the crater rim.

To see evidence of the impact origin of the basin, one can visit (by boat) the numerous low cliffs that make up the shoreline, where the highly-fractured rock is visible. More obvious evidence comes from deep cores taken from the center of the lake, which is about 100 m deep. Beneath the water are 100 m of lake and glacial sediments, which are underlain by about 350 m of impact breccia. As at Brent, the breccia overlies 200 m of greatly fractured rock.

Further evidence of the impact origin of West Hawk Lake comes from a gravity survey, which shows a negative anomaly that results from the symmetrical basin of brecciated rock. Shock-induced fractures in quartz grains in the drilling cores are also present.

Access: West Hawk Lake is easily reached by road from Winnipeg by driving east on the Trans-Canada Highway (#1) about 150 km. The lake is in the Whiteshell Provincial Park and there is a campground along its shore.

References

Beals, C. S. and Halliday, I., 1965, Impact craters of the Earth and Moon. *J. Royal Astron. Soc. of Canada*, *59*, 199–216.

Halliday, I. and Griffin, A., 1963, Evidence in support of a meteoritic origin for West Hawk Lake, Manitoba, Canada. *J. Geophys. Res.*, *68*, 5297.

4

Impact structures of Latin America

Araguainha Dome

Brazil

Lat/Long: S16° 46′, W52° 59′
Diameter: 40 km
Age: <249 Ma
Condition: Eroded

The giant Araguainha Dome is Brazil's largest and most complex impact structure. It lies in the Parana Basin astraddle the boundary between the states of Mato Grosso and Goias. The country rocks are Paleozoic sediments, and the dome was thought at first to be a 'blister' in the sedimentary rocks, pushed up by a volcanic intrusion associated with Jurassic–Cretaceous volcanism. How-

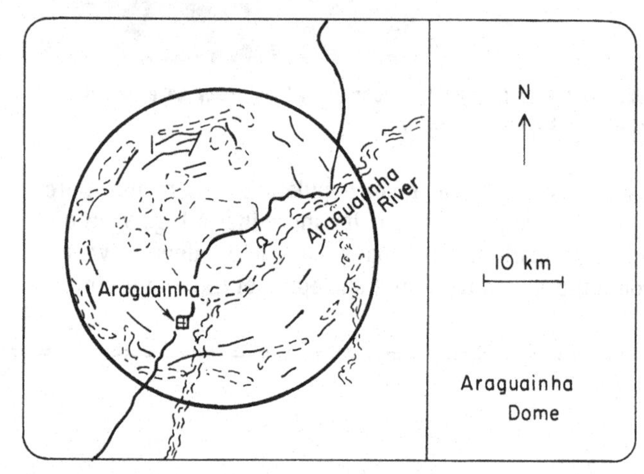

ARAGUAINHA DOME: Sketch map of the Araguainha Dome, showing the dome's structure at the surface (after Crosta).

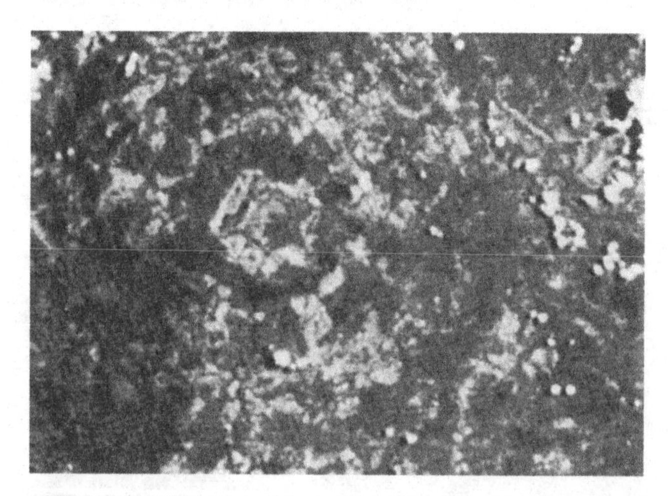

ARAGUAINHA DOME: View of Araguainha Dome from space (courtesy of NASA).

ever, its remarkable similarity in appearance on LANDSAT images to ringed craters on the moon, as well as to other large terrestrial impact basins, led scientists to consider a possible meteoritic origin, which has now been well-established.

Principal evidence about the origin of Araguainha Dome comes from the breccia found near its center and the shock lamellae detected in its grains of quartz, as well as shatter cones and other forms of shock metamorphism. The Dome's structure includes a 6 km central uplift of basement granite with heavily faulted and deformed morphology. Surrounding this is a raised rim with circular faults that have formed graben arcs. Its multi-ring shape is typical for impact structures of its size.

Access: Araguainha Dome is located 650 km west of Brasilia on the edge of the Mato Grosso. The Araguainha River cuts through the center of the structure and the village of Araguainha is a few kilometers to the west.

References

Crosta, A. P., 1987, Impact structures in Brazil, in *Research in Terrestrial Impact Structures*, ed. J. Pohl, pp. 30–37. Vieweg and Sohn, Braunschweig.

Dietz, R. S. and French, B. M., 1981, Two probable astroblemes in Brazil. *Nature*, *244*, 561–562.

Campo Del Cielo Craters

Argentina

Lat/Long: S27° 38′, W61° 42′
Diameter: 0.115–0.020 km
Age: <0.004 Ma
Condition: Slightly eroded, largely brush-covered

The Campo del Cielo area in northern Argentina has long been known to be a site of meteorites and craters. There is an ancient tradition that a large rock fell from the sky here, reflected in the aboriginal name for the area, Piguem Nonralta ('Field of the Sky'), which was translated into the Spanish 'Campo del Cielo' by the conquistadors. The fabled giant meteorite known to the native peoples was called the Meson de Fierro by the Spanish, who brought back pieces of it from five expeditions to the area, the first in 1576 and the last in 1783. The location of the giant Meson de Fierro was then lost and it has been the object of searches by treasure hunters ever since. Smaller meteorites, up to a few thousand kilograms in weight,

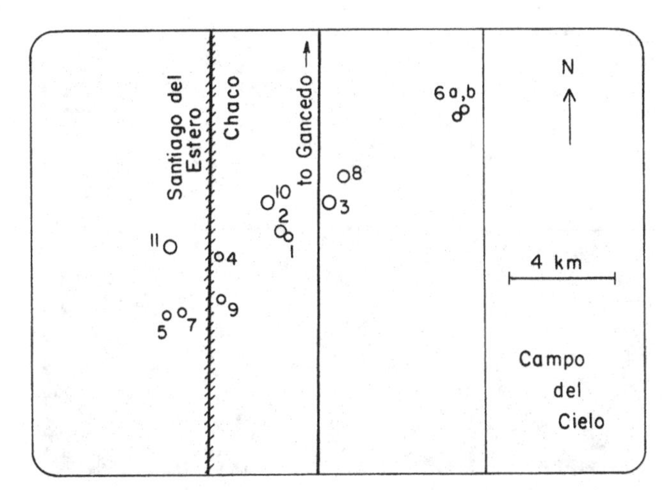

CAMPO DEL CIELO: Location map of the craters at Campo del Cielo (after Cassidy).

have been recovered at Campo del Cielo and are located in many museums and collections.

In addition to many meteorites found at the site, there are 12 small craters, with other craters still being searched for and examined. All of the craters are small and thus many have had much of their meteoritic material found within their boundaries. They lie on a rather narrow line oriented roughly northeast-southwest, along which most of the large meteorites were also found. The crater field is unusually elongated, with a linear extent of 17 km, which can be contrasted with those of the Henbury Craters (0.7 km) and the Sikhote-Alin group (1.2 km). The largest crater has a rim-to-rim diameter of 100 m and the smallest is only about 20 m across. Several of them have been given proper names in addition to numbers. The following description of the more notable craters is based primarily on the article (referenced below) by Cassidy, who led a multi-year international investigation of the crater field beginning in 1961.

Crater 1

This crater is named 'Hoyo de la Canada'. It lies near the center of the strewn field, at the southwest boundary of the area of the most densely-dispersed meteorites. It is elliptical in shape, possibly because of being formed by two adjacent meteorite fragments. The major dimension is 105 m from rim to rim. A shallow gully found in the rim gives it its name. Trenching in the 1960s turned up many miscellaneous objects in the central layers of sediment, including meteorite fragments, meteoritic shale, bones and charcoal. The crater is presently 2 m deep at its deepest.

Crater 2

Named 'Hoyo Rubin de Cellis' after the explorer Navy Lt. don Miguel Rubin de Cellis, who led an expedition to the

CAMPO DEL CIELO: Photograph of Craters 1 and 2, Hoyo Rubin de Celis (courtesy of W. A. Cassidy).

CAMPO DEL CIELO: Photograph of Crater 3 (courtesy of W. A. Cassidy).

area in 1783 to search for the fabled Meson de Fierro, Crater 2 has been studied extensively by the Cassidy group. It has a diameter of 70 m and is the deepest crater (5 m) and probably the least eroded. Cassidy has speculated that it may be the first identified meteorite crater in the world, as he believes that it is the depression known for centuries to the native peoples as the 'Pool of the Sky'. An extensive radial trench through the crater and its rim showed many features common to impact craters of its size, including upthrust of the rim by about 0.5 m and inversion of the stratigraphy outside of the rim. Drilling at the center showed the presence of 'clay breccia' at depths to 15 m below the present floor. Meteorite fragments are found at depths of 4 to 9 m. A charcoal stump was discovered under the ejecta outside the rim and carbon- 14 dating gives it an age of 5800 years, which is probably the age of the craters.

Crater 3

Called 'Laguna Negra' because of the lake that filled it when it was first studied in 1923, Crater 3 is the largest crater, with a major dimension of 115 m. It is, however, quite shallow, only about 2 m deep at the center. A fossil jaw of a dog was found in the sediments of the crater and the species was identified as being the same as a wild dog now limited to areas of Colombia. It is interesting that this mammal was present in Argentina somewhat less than 5800 years ago but is now unknown there.

Crater 4

Probably identified as the crater 'Hoyo Aislado', Crater 4 is 85 m in diameter and 1.5 m deep. Large trees line the crater and its contours have been considerable altered by cattle, who have descended into it to drink water after rains.

Crater 5

Crater 5 is shallow with an ill-defined rim. The floor is about 45 m in diameter. It defines the southwest end of the crater field.

Craters 6a and 6b

These twin craters share a common east-west rim. They were discovered by Carmen Sosa, a local resident, in the 1930s and then were lost until he rediscovered them in 1963 after searching on foot for three weeks. The larger, 6a, has a diameter of 35 m, while the smaller is 20 m across. They define the northeast end of the crater field.

Crater 7

Crater 7 is elliptical in outline with rim-to-rim dimensions of 96 m × 74 m.

Crater 8

Also elliptical, Crater 8 has dimensions of 46 m × 28 m and a depth of only a fraction of a meter. A total of 25 meteorites were recovered from the crater, all but one from the sediments in its floor.

Crater 9

Named 'La Perdida', Crater 9 was discovered in 1963 by a local resident while hunting iguanas. A 1965 magnetometer map of the floor of La Perdida showed about a dozen magnetic anomalies and at the site of the deepest of these, meteoriticists recovered a 1.5 metric ton meteorite.

Craters 10 and 11

Both of these recently-discovered craters lie to the north of the line that defines the average crater position.

Access: The Campo del Cielo Craters are about 1000 km northwest of Buenos Aires. They inhabit the relatively hot, arid plain called the Gran Chaco. The crater field straddles the boundary between the states of Chaco and Santiago del Estero and is just northwest of the town of Chorotis. It is 15 km directly south of the settlement of Gancedo.

References

Cassidy, W. A., 1967, Meteorite field studies at Campo del Cielo. *Sky and Telescope*, *34*, 2–8.

Cassidy, W. A., Villar, L., Bunch, T., Kohman, T. and Milton, D., 1965, Meteorites and Craters of Campo del Cielo, Argentina. *Science*, *149*, 1055–1064.

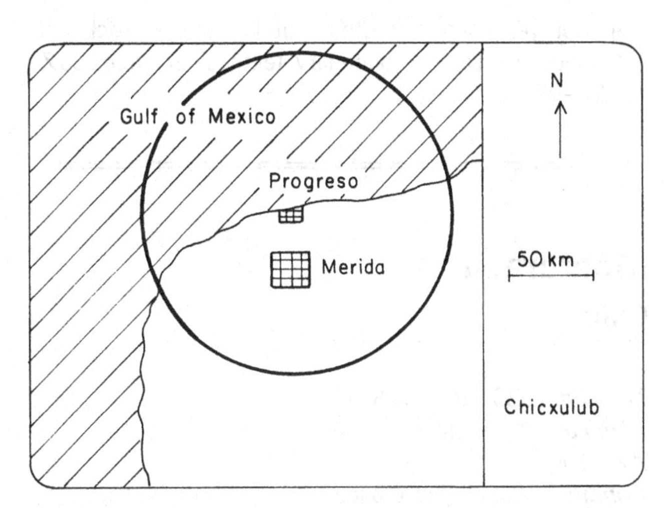

CHICXULUB: Location map of the subterranian Chicxulub structure (after Hildebrand *et al.*).

Chicxulub

Mexico

Lat/Long: N21° 24′, W89° 31′
Diameter: 180 km
Age: 65 Ma
Condition: Buried

The Chicxulub Crater experienced a somewhat backwards discovery history. For most impact structures, the crater was found first and then the ejecta were identified. For Chicxulub, however, the ejecta were identified more than 10 years before the crater was found. The probable ejecta are the famous Cretaceous–Tertiary (K–T) boundary layer that was first identified as the result of an extraterrestrial impact by the fact that it shows a very high iridium abundance, this element and its close relatives being rare in terrestrial surface materials compared with meteorites. There are also other types of evidence that the layer was impact-produced: clay spherules that were originally impact melt droplets (similar to tektites) are found in the sediments and shocked quartz grains have been found in the K–T layer throughout the world. Furthermore, the size of the layer and its fragments increases radially towards a position somewhere in the Caribbean (the fine, iridium-rich dust layer is world-wide, but the thick ejecta layer is only found in eastern North America and the Caribbean). Another clue comes from the chemical composition of the tektite glass, which indicates the presence of both andesitic continental rock and carbonates, suggesting that the projectile hit the Earth somewhere near a coastline. To add to this suggestion, geologists have found evidence of a tsunami that caused huge waves to leave debris at the paleoshorelines between North and South America. All of these events have been dated and found to have occurred at the same time, 65 million years ago, which is also the date of the Cretaceous-Tertiary extinction.

The Chicxulub Crater was identified in 1990, when one of the students of the ejecta learned of a sub-surface, circular structure in the Yucatan peninsula that had been found by drilling activity by the Mexican national petroleum company. Subsequent examination of the cores presented strong evidence for the presence of a large impact structure; there were impact melt rocks, impact breccia, and grains of shocked quartz. Furthermore, the composition of the melt rocks was found to be similar to that of the tektite glass. An age for the crater was determined in 1992 by analyzing the impact glass from the drill cores. It is 65 million years, the same as the age of the tektites and of the K–T boundary. This last measurement has largely put to rest the continuing doubt that many people had about the impact origin of the boundary event.

Access: The crater is buried under almost a kilometer of sediments. However, one can visit the site, which occupies part of the northwest corner of the Yucatan peninsula, including the city of Merida. From the air, one can see an arc of sinkholes that outline the structure's edge.

References

Hildebrand, A., Penfield, G. T., Kring, D. A., Pilkington, M., Camargo, A., Jacobsen, S. B. and Boynton, W., 1991, Chicxulub Crater: a possible Cretaceous/Tertiary boundary impact crater on the Yucatan Peninsula, Mexico., *Geology*, *19*, 867–871.

Swisher, C. C. *et al.*, 1992, Coeval^{40}A/^{39} A ages of 65.0

million years ago from Chicxulub Crater melt rock and Cretaceous–Tertiary boundary tektites. *Science, 257,* 964–958.

Monturaqui
Chile

Lat/Long: S23° 56′, W68° 17′
Diameter: 0.46 km
Age: 1 Ma
Condition: Somewhat eroded

The Monturaqui Crater was discovered in 1962 by examination of aerial photographs and confirmed three years later by field work that identified meteoritic iron shale and impact glass at the site. The iron shale was found on the outer rim of the crater and the impact glass material was abundant on the south and southeast flanks of the crater. The impactites have shocked and unshocked mineral grains and rock fragments, as well as Fe–Ni rich spherules, all bound in glass.

The crater lies in the rugged desert hills of the Monturaqui mountain range and is crossed by a long-established trade route (a llama trail) connecting the Pacific coast with San Pedro de Atacama and northern Argentina. The present drainage system skirts the crater. Its lowest point is marked by a bright deposit of clay and silt.

Access: The crater is in the high Atacama Desert near the Chile–Bolivia–Argentina border. It is 120 km directly south of the town of San Pedro de Atacama and about 30 km south of the Salar de Atacama.

References

Bunch, T. and Cassidy, W., 1972, Petrographic and electron microprobe study of the Monturaqui impactite. *Contr. Minerol. Petrol., 36,* 95–112.

Sanchez, J. and Cassidy, W., 1966; A previously undescribed meteorite crater in Chile. *J. Geophys. Res., 71,* 4891–4895.

Riachao Ring
Brazil

Lat/Long: S7° 46′, W46° 39′
Diameter: 4 km
Age: <200 Ma
Condition: Eroded

Discovered by Apollo mission astronauts, the Riachao structure is typical of its size. There is a central uplift and an outer graben, with both radial and concentric faults. The country rocks are sandstones. There are occurrences of polymict breccias with microfractures visible in quartz grains.

Access: The Riachao Ring is near the village of Riachao, 60 km west of Balsas, which is accessible by road from the east.

References

Crosta, A. P., 1987, Impact structures in Brazil, in *Research in Terrestrial Impact Structures,* ed. J. Pohl, pp. 30–37. Vieweg and Sohn, Braunschweig.

McHone, J., 1979, Riachao Ring, Brazil: a possible meteorite crater discovered by the Apollo Astronauts. *Apollo-Soiuz Test Proj., Rpt., 11,* 193–202.

Rio Cuarto Craters
Argentina

Lat/Long: S32° 52′, W64° 14′
Diameter: 4.5 × 1.1 km
Age: <0.01 Ma
Condition: Slightly eroded

The remarkable Rio Cuarto Craters were first noticed in 1990 by an airplane pilot who noticed an odd alignment of depressions in the otherwise smooth pampas of north-central Argentina. Exploration on the ground in the following year demonstrated that these, together with several others, made up a series of at least ten oblong meteorite craters, ranging in size from a few kilometers down to about 250 m along their long axis. All are aligned in parallel in a northeat–southwest direction and they span a distance of about 30 km. The larger depressions have rugged edges and therefore have not been subjected

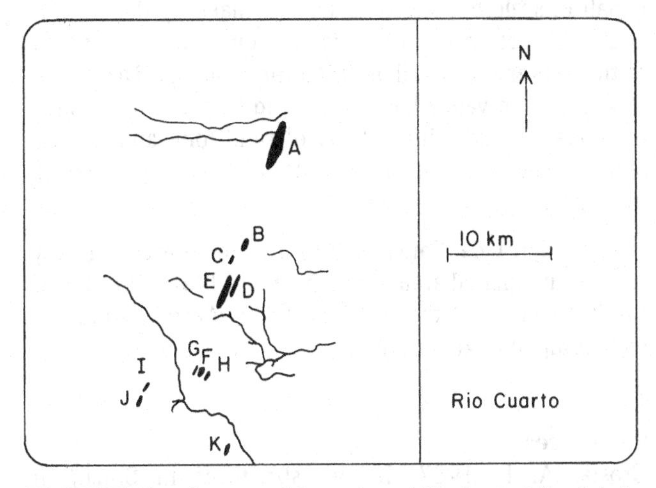

RIO CUARTO: Location map of the craters at Rio Cuarto (after Schultz and Lianza).

to farming and its attendant erasure of details, but some possible smaller structures exist which have been tilled and which show up only as disturbances in the drainage or as possessing unusual patterns of plant growth.

The rims of the largest structures are well-defined along the edges but poorly-defined at each end. The intermediate-sized craters retain more perfect rims. In general the rims are 3–7 m above the plains and the floors of the larger craters are about 10 m below the rims.

The morphology of the crater field strongly resembles experiments with high-velocity glancing impacts made in the laboratory. Modeling and scaling from experiments suggests that the Rio Cuarto craters were formed by a low-angle impact of an asteroid approximately 150 m in diameter. The projectile entered from the northeast and possibly broke into fragments, with the largest fragment scouring out the first groove and smaller fragments continuing on downrange to calve smaller valleys. Possibly there was only one fragment at first, which collided with the ground and then portions of it ricocheted on to form further craters (laboratory experiments suggest that it could have suffered 'decapitation' in this way).

Two kinds of material have been recovered from the craters that indicate its meteoritic origin. There are many examples of tektite-like impact glass, which strongly resemble in composition and physical structure other known impact glasses. They show flow patterns, small vesicles and quartz inclusions and are often greenish in color and twisted and elongated in shape. The other kind of material is represented by two small chondritic meteorites that were found in the craters. One was enveloped in a shell of impactite-like material, indicating that it was probably a small fragment of the incoming meteorite that was relatively unaltered but was covered by a glassy mixture of material from the projectile and the country rock.

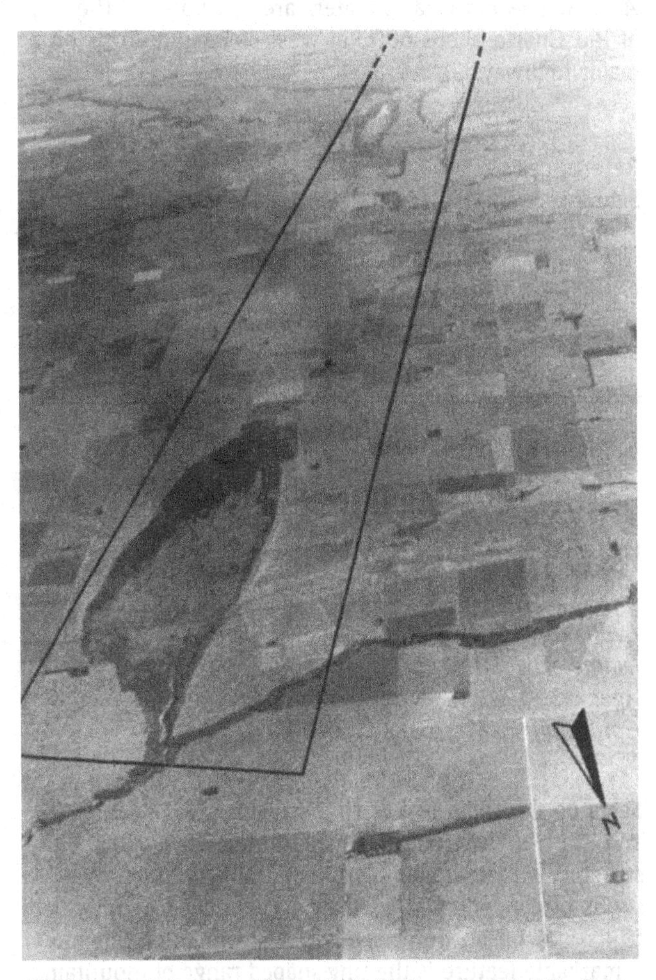

RIO CUARTO: Aerial view of Crater A (the 'Northern Basin') at Rio Cuarto. The long, narrow area occupied by the craters is marked and craters E and D are conspicuous in the distance (photo courtesy of P. Schultz and R. Lianza).

RIO CUARTO: Aerial view of craters D (right) and E (left) at Rio Cuarto, looking north-northeast. Craters A, B, and C can be seen dimly in the distance (photo courtesy of P. Schultz and R. Lianza).

Access: The Rio Cuarto craters are just north of the city of Rio Cuarto about 600 km west of Buenos Aires on a major highway.

References

Schultz, P. and Lianza, R., 1992, Recent grazing impacts on the Earth recorded in the Rio Cuarto crater field, Argentina. *Nature*, *355*, 234–237.

Schultz, P. and Beatty, J. K., 1992, Teardrops on the pampas. *Sky and Telescope*, *83*, 387–392.

Serra Da Cangalha

Brazil

Lat/Long: S8° 05′, W46° 52′
Diameter: 1.2 km
Age: <300 Ma
Condition: Exposed and eroded

Serra da Cangalha makes up a structure similar in shape and state of erosion to the better-known Gosses Bluff structure in Australia. Morphologically, it consists of a series of concentric rings that are delineated by circular faults and by an outer semi-circular rim. But the most remarkable feature is the ring-shaped range of mountains (small in scale but spectacular) that make up the core. A nearly-perfect circle 250 m high surrounds a hidden valley at the center – a Brazilian wilderness Shangri-La.

SERRA DA CANGALHA: View from space of Serra da Cangalha. The central ring of hills is conspicuous and the outer ring is at the edges of this view (courtesy of NASA).

The hidden valley has been found to house many shatter cones in its uplifted beds of sandstone and breccia and the quartz grains of the breccia show intense shock fracturing.

Access: Serra da Cangalha lies in the forest of a very sparsely-populated area of north-central Brazil. It lies near the headwaters of the Parnaiba River. Access requires a well-equipped expedition.

References

Crosta, A. P. 1987, Impact structures in Brazil, in *Research in Terrestrial Impact Structures*, ed. J. Pohl, pp. 30–37. Vieweg and Sohn, Braunschweig.

Dietz, R. S. and French, B. M., 1981, Two probable astroblemes in Brazil, *Nature*, *244*, 561–562.

Vargeao Dome

Brazil

Lat/Long: S26° 50′, W52° 07′
Diameter: 12 km
Age: <70 Ma
Condition: Eroded

The Vargeao Dome is an eroded impact structure in southern Brazil, first identified in 1978. It is conspicuous because of its arrangement of circular, concentric fractures. There is a considerable amount of impact breccia and a central uplift, with beds about 500 m above their normal location. Planar features are detected in quartz grains.

Access: The structure is in the state of Santa Catarina, northeast of the city of Xapeco.

Reference

Crosta, A. P., 1987, Impact structures in Brazil, in *Research in Terrestial Impact Structures*, ed. J. Pohl, pp. 30–37. Vieweg and Sohn Braunschweig.

5

Impact structures of Australia

Acraman

South Australia

Lat/Long: S32° 01′, E135° 27′
Diameter: 160 km
Age: >570 Ma
Condition: Highly eroded, small amount of exposure

The Acraman Impact Structure is Australia's largest. It was discovered in 1986 and has since been of considerable interest because of the extensive blanket of ejecta that was produced and that can still be traced out to distances as large as 500 km. Ejecta have been traced over an area of 20 000 km^2. Particularly nice examples of the Acraman ejecta are found, for example, in Flinders National Park, 300 km east of Acraman, in the Bunyeroo Formation, which otherwise consists of Proterozoic marine shales. The ejecta layer includes angular clasts ranging in size from boulders to fine sand, in which are mixed microscopic spherules that strongly resemble microtectites, though they have been metamorphosed over the age of the material from their initial glass composition to more stable phases. The fragments of rock in the ejecta layer show shock lamellae, small shatter cones, and anomalous abundances of iridium and other precious metals.

The center of the crater site is presently occupied by a seasonal lake, Lake Acraman. There is a central uplift area of shattered and shock-deformed rock (the country rock is Yardea Dacite) and shatter cones can be found in a small outcrop on an island near the center of the present

ACRAMAN: Lake Acraman is the shallow central lake of the Acraman basin. The island with exposed shatter cones is near the distant shore in this photograph, which was taken looking west.

lake. The central uplift is about 10 km in diameter and is surrounded by a general depression that is 35 km in diameter. The very large size assigned to the structure (160 km) is the result of the presence of a ring of circular valleys at a distance of about 45 km from the center and a more distant series of arcuate features at a radial distance of about 80 km. Some geologists, however, consider the true crater diameter to have been much smaller, perhaps about the size of the exisiting 35-km depression.

Access: Lake Acraman is accessible by dirt and gravel roads. Drive west on the Eyre Highway 300 km from Port Augusta and turn right at Minnipa. The country road proceeds north to Yardea Homestead, 75 km from the Highway. Turn left onto a track that passes Chinaman Dam. The lake is reached in about 50 km. The lake lies to the west of Jumpuppy Hill. Emus and wombats are abundant in the area.

References

Gostin, V., Haines, P., Jenkins, R., Compston, W. and Williams, G., 1986, Impact ejecta horizon within Late Precambrian shales, Adelaide geosyncline, South Australia. *Science*, *233*, 198–200.

Wallace, M., Gostin, V. and Keays, R., 1990, Spherules and shard-like clasts from the Late Proterozoic Acraman impact ejecta horizon, South Australia. *Meteoritics*, *25*, 161–165.

Williams, G., 1986, The Acraman impact structure: source of ejecta in Late Precambrian shales, South Australia. *Science*, *233*, 200–203.

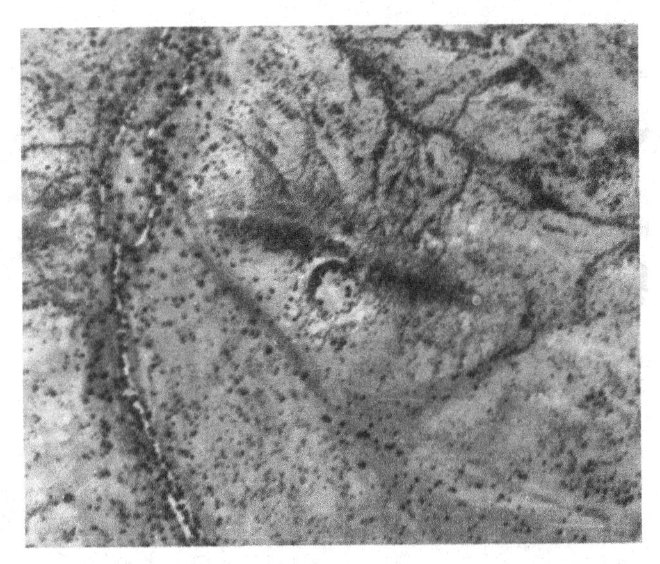

BOXHOLE: Aereal view of the crater and its surroundings (north is up) (courtesy of Division of National Mapping, Australia).

BOXHOLE: View of the east rim of the Boxhole crater from the ridge to the north. The crater floor is to the right of the center of this photo.

Boxhole Crater

Northern Territory

Lat/Long: S22° 37′, E135° 12′
Diameter: 0.185 km
Age: 0.005 Ma
Condition: Fresh, mildly eroded

The Boxhole crater is a small, relatively-recently-formed crater northeast of Alice Springs. It lies on the south side of an east–west-running ridge approximately 10 m high. The crater occupies a part of a gently sloping alluvial plane. It is 185 m across and its rim is raised 3 to 5 m above the surrounding ground level. The floor of the crater is largely bare of trees and brush, except for a single euca-

BOXHOLE: The floor of the Boxhole crater looking east from the west rim.

lyptus. The floor is covered with fine-textured dry lake sediments. The rim is covered with small brushy trees.

Many fragments of the impacting meteorite have been collected in the surroundings. It is a medium octahedrite. Impact-melted meteoritic spherules have been recovered from the soil, but they are very rare. Meteoritic shale balls have been recovered, primarily from the north rim and the ridge to the north.

An ejecta blanket has been mapped around the crater and it is found to be thickest at the south rim and absent at the north rim, leading to the conclusion that the meteorite entered from the north.

Access: The Boxhole Crater lies 1 km north of the Dnieper Station, which is reached by a track that runs 30 km north from the Plenty Highway, leaving it about 192 km east of the junction of the highway with the Stuart Highway. The distance by road from Alice Springs is about 270 km.

References

Hodge, P. and Wright, F., 1973, Particles around the Boxhole meteorite crater. *Meteoritics*, *8*, 315–320.

Madigan, C. T., 1940, The Boxhole meteoritic iron, Central Australia. *Mineralog. Mag.*, *25*, 481–483.

Milton, D. J., 1968, The Boxhole meteorite crater. *US Geol. Surv. Prof. Pap.*, 599-C, 1–23.

Shoemaker, E. M., Roddy, D. J., Shoemaker, C. S. and Roddy, J. K., 1987, The Boxhole meteorite crater, Northern Territory, Australia. *Lunar Planet. Sci.*, *19*, 1081–1082.

Connolly Basin

Western Australia

Lat/Long: S23° 32′, E124° 45′
Diameter: 9 km
Age: <60 Ma
Condition: Eroded, partly exposed

Connolly Basin is a circular depression in the Gibson Desert. Drainage in the area shows a radial pattern towards the center of the basin. The basin floor is flat, covered primarily with spiny spinafex. There is a rim surrounding the basin, about 30 m above the general level. Seismic measures show that the structure is shallow and circularly-symmetric. It is filled with post-crater sediments.

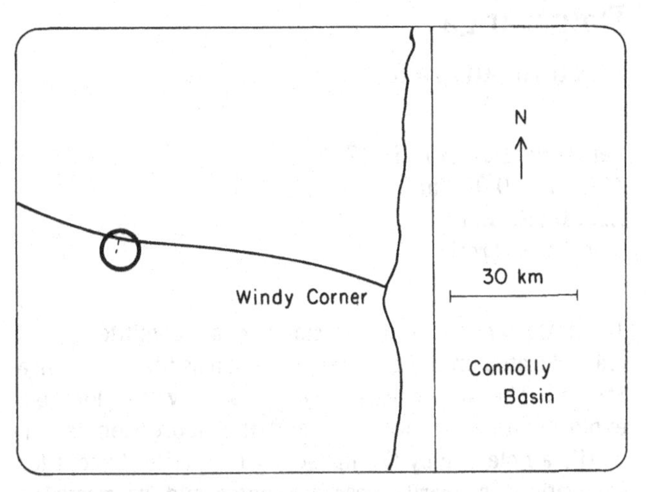

CONNOLLY BASIN: Location map for the Connolly Basin. Roads indicated are rough desert tracks.

CONNOLLY BASIN: The central area of the Connolly Basin looking south from the northern 'rim' towards the outcrops of rock identified as the central uplift.

The central uplift consists of a 1 km diameter ring of puckered, highly-deformed sandstone. Quartz grains show shock effects. Inside the sandstone ring are deformed shales and siltstones.

Access: The Connolly Basin can be reached from the town of Wiluna on desert roads by proceeding east on the 'Gunbarrel Highway' (very rough washboarding) to Everard Junction, then north to Windy Corner. An east–west track passes through the north part of the basin.

Beware: On at least one commercially-available map of the area, the basin is incorrectly placed several kilometers to the west of its correct position.

Reference

Shoemaker, E. and Shoemaker, C., 1988, Impact structures of Australia. *Lunar Planet. Sci.*, *19*, 1079–1080.

Dalgaranga

Western Australia

Lat/Long: S27° 43′, E117° 5′
Diameter: 0.02 km
Age: 0.025 Ma
Condition: Fresh

The Dalgaranga Crater is a small, relatively-little-eroded crater located on the giant ranch known as the Dalgaranga Station. The crater is too small for any considerable explosion to have occurred, and the depression is primarily a hole dug by the impacting meteorite. Except for excavations primarily near its center and its northern floor, the crater is relatively unchanged, the area being extremely arid. Breccia can be found in the crater floor and along the south wall. On the north rim the layers of the country granite are tipped up vertically and in some places entirely over. On the south rim the layers are tipped up more gently. A ray of ejecta, not unlike the rays of lunar craters, extends to the east, reaching some 40 m from the southeast crater rim.

DALGARANGA: South rim of the Dalgaranga crater. The bottom of the crater is out of the picture to the right.

DALGARANGA: The outer rim of the Dalgaranga crater showing upturned strata and ejecta. Eugene Shoemaker points to the north rim.

DALGARANGA: Upturned layers of granite at the northeast rim of the Dalgaranga crater.

The meteorite that caused the Dalgaranga Crater was a mesosiderite, hundreds of fragments of which have been collected, many from an area a few tens of meters to the southwest of the crater.

Access: Dalgaranga Station can be reached on a good gravel road in about 30 minutes from Yalgoo. The crater is 35 km from the station by track; permission to visit should be obtained, as it is on private land.

Reference

Nininger, H. H. and Huss, G., 1960, The unique meteorite crater at Dalgaranga, Western Australia. *Mineralog. Mag.*, *32*, 619–639.

Darwin Crater

Tasmania

Lat/Long: S42° 18′, E145° 40′
Diameter: 1 km
Age: 0.73 Ma
Condition: Heavily eroded and densely vegetated

For nearly 70 years samples of natural, anomalous glass found near Mt. Darwin in western Tasmania remained a puzzle. Chemically they were similar to tektites, but otherwise more like impact glass, though no impact crater was evident. The Darwin glass is found in an area of rain forest that is extremely dense and any search for either more glass (to delineate the strewn field) or a crater had to wait until a road was cut into the region.

In 1972 the hydroelectric utility penetrated into the country east of Mt. Darwin and the area became more accessible. Many new sites of Darwin glass were found and a crater-form structure was located in a valley about 2 km from the road. Though having no rim or other obvious crater topography, the crater was eventually confirmed by a gravity survey, drilling, and by circumstantial evidence from the Darwin glass, the fragments of which monotonically increase in mean size towards the crater. The strewn field is asymmetrical in shape, with most of the glass located north, west and south of the crater. It has been estimated that there must be several hundred tons of Darwin glass in total.

Drilling at the center of the crater showed a bed of lake sediments, underlain by clay, rock fragments and sand. At 100 meters was a sand layer that contained grains of Darwin glass and of lechatelierite. It is estimated, from the gravity survey, that a relatively undisturbed limestone layer at the floor of the crater lies about 200 m below the present surface.

Access: The Darwin Crater is located in hilly, densely vegetated country 26 km south of Queenstown. It is 2 km east of the Franklin Track.

References

Fudali, R. and Ford, R., 1979, Darwin glass and Darwin Crater: a progress report. *Meteoritics, 14*, 283–296.

Meisel, T. and Koeberl, C., 1988, Geochemical studies of impact glass from the Darwin Crater, Tasmania. *Meteoritics, 23*, 289–290.

Goat Paddock

Western Australia

Lat/Long: S18° 20′, E126° 40′
Diameter: 5 km
Age: 55 Ma
Condition: Eroded

Goat Paddock is a smooth-floored circular basin in a treeless area of the Kimberley District of Western Australia.

GOAT PADDOCK: Aerial view of the Goat Paddock crater, showing its eroded rim and its flat, sediment-filled floor.

Its principal feature is the rather steep walls of the basin, which delineate the structure. The geology of the structure is revealed by the exposures of rock found in several deep canyons that cut into these walls. Beds of the country rock are tilted up away from the crater and in some locations they are overturned. Shatter cones have been recovered and breccias and impact melt rocks have been located under the crater, as well as in the exposed rocks of the walls. Although the crater is large enough to be a complex type, no evidence of a central uplift has been reported.

Access: Goat Paddock is 50 km north of the Great Northern Highway and about 110 km west of the city of Halls Creek. No roads reach the structure but it can be reached overland. It is just north of the Mueller Ranges and south of the O'Donnell River.

References

Harms, J., Milton, D., Ferguson, J., Gilbert, D., Harris, W. and Goleby, B., 1980, Goat Paddock cryptoexplosion Crater, Western Australia. *Science, 286*, 704–706.

Shoemaker, E. and Shoemaker, C., 1987, Impact structures of Australia. *Lunar Planet. Sci., 19*, 1079–1080.

Gosses Bluff

Northern Territory

Lat/Long: S23° 50′, E132° 19′
Diameter: 22 km
Age: 142 Ma
Condition: Highly eroded

In an area of Australia that is famous for its remarkable rock formations (e.g., Ayers Rock, Stanley Chasm, Mount Olga), Gosses Bluff is one of the most remarkable. Standing alone on a nearly flat plain, it rises abruptly to a flat top, looking from the distance like a lone, circular mesa. However, from the air one sees that the mountain is completely hollow. It is a nearly-perfectly-circular small mountain range, with a flat, featureless floor in the center. Breached on the east side by a winding dry river bed, the mountain is a ring-shaped wall of rock.

Geological exploration in the late 1960s showed that the conspicuous ring of Gosses Bluff is the inner part of a highly-eroded impact structure. The ring is the skeletal remains of the central uplift, the center of which, being made of softer rock, has eroded away. Almost all of the rest of the structure is also gone; only a few remnants of the crater rim are found far outside the mountain ring. For instance, the rock formation known as the Wizard of Oz is a fragment of crater rim in which may be seen the uptilted beds of the original outer crater rim.

The center of Gosses Bluff is virtually paved with shatter cones. The various Ordovician sandstones that were the target rocks made particularly clear shatter cones, which are found not only all over the floor of the central valley, but also at various places in the cliffs and on the tops of the ridges. Monomict breccia is also abundant in the structure.

Access: Gosses Bluff is west of Alice Springs, from which it can be reached on mostly unpaved roads. The road to

GOSSES BLUFF: Aerial view of Gosses Bluff, showing the form of the central ring of mountains (photo courtesy of Daniel Milton).

GOSSES BLUFF: A road cuts through the Gosses Bluff ring of mountains at the east, where there is an intermitent stream outlet.

GOSSES BLUFF: Gosses Bluff from the plains to the north.

GOSSES BLUFF: The floor of the central depression of Gosses Bluff, looking northeast towards the inner wall.

GOSSES BLUFF: Rock fragments on the floor of the central depression of Gosses Bluff, including shatter-coning patterns.

GOSSES BLUFF: Near the top of the north rim of the Gosses Bluff inner ring, showing examples of impact breccia. A fragment of sandstone in the breccia near the center of the photograph shows shatter-coning.

Hermannsburg (78 km) is followed and then one proceeds northwest on narrow roads 70 km to a turn-off to the left, a point where there is a good view of Gosses Bluff to the south.

References

Milton, D. J. *et al.*, 1972, Gosses Bluff impact structure, Australia. *Science, 175,* 1199–1207.

Milton, D. J. and Sutter, J. F., 1987, Revised age for the Gosses Bluff impact structure, Northern Territory, Australia, based on $^{40}Ar/^{39}Ar$ dating. *Meteoritics, 22,* 281–289.

Henbury Craters

Northern Territory

Lat/Long: S24° 35′, E133° 09′
Diameters: 0.157–0.006 km
Age: <0.005 Ma
Condition: Exposed, slightly eroded

The Henbury Crater Field includes several meteorite craters of young age, arranged in an ellipse, with the largest craters to the north and the smallest at the south end of the pattern. The largest crater is highly elliptical and is probably the result of the impact of two fragments of the incoming meteorite, which probably broke up in the atmosphere. Early maps include some craters that do not seem to be genuine and some newly-recognized craters have been added in the years since the original studies in the 1930s. The total number of craters that is generally agreed-upon is 11, counting the 'main crater' as one.

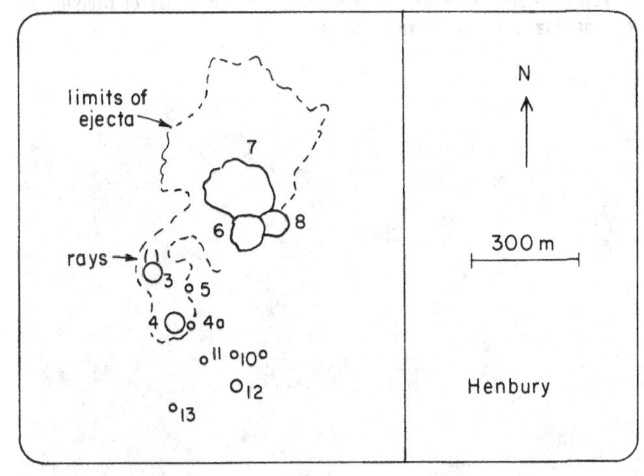

HENBURY: Identification map of the confirmed craters at Henbury.

According to the field work of Milton, Roddy and E. and C. Shoemaker, the characteristics of the craters are as follows (the numbering is from Alderman's original listing, which included some craters that are now considered doubtful):

Crater 3

An intermediate-sized crater, Crater 3 is distinctive because of its remarkable system of 'rays', which consist of ejected fragments of sandstone that were emitted to form looped rays similar to those found around fresh craters on the Moon. The mean diameter of the crater (rim crest to rim crest) is 65 m. Its well-preserved rim stands about 0.5 m above the surroundings and its floor lies 4 m below the rim.

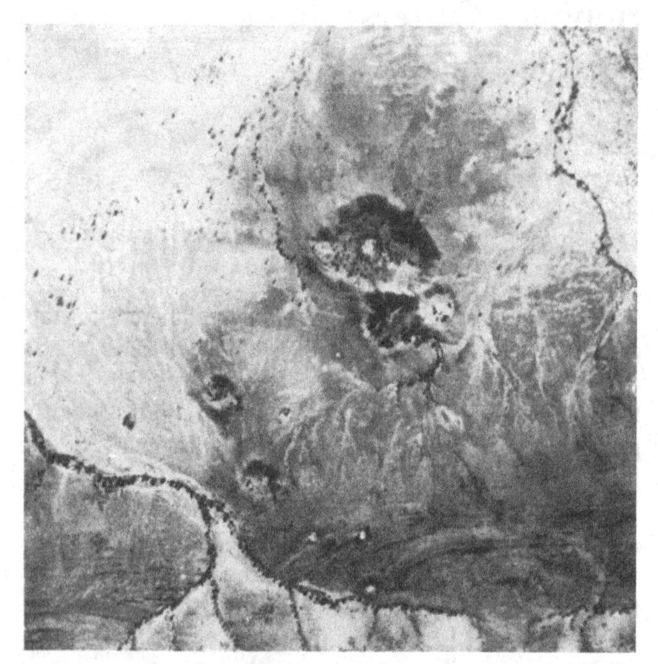

HENBURY: Aerial view of the Henbury craters (courtesy of Division of National Mapping, Canberra Australia).

HENBURY: A low-altitude aerial view of the western portion of the Henbury crater field. Craters 1 and 2 are questionable.

Crater 4

Almost but not quite an identical twin of Crater 3, number 4 lies immediately to its south. Its mean diameter is 70 m, its rim averages about 1 m high and its floor lies 4 m below the rim. It has rays of ejecta like those of Crater 3, one of which extends westward from the rim's outer base to a distance of 75 m.

Crater 4a

This small crater was first identified by D. Milton in 1963 and confirmed by E. Shoemaker in 1987. It is about 20 m in diameter and shares a rim with Crater 4. It is inconspicuous because of being cut by the stream that feeds into the water crater.

Crater 5

Very little remains of Crater 5, a small crater largely destroyed by erosion. It is 18 m in diameter and has very little depth, with a barely detectable rim. Its floor is flat and smooth, filled with recent sediments.

Crater 6 (the 'Water Crater')

Called the Water Crater because its fine-textured sediment floor holds rainwater better than the surrounding terrain and because a stream feeds into it, Crater 6 is one of the three largest craters. It is filled with water (in season), mud, footprints and trees. The trees make it easy to spot from a distance. The height of the rim is about 2 m in the south, but nearly 4 m on the north, where it shares its rim with Crater 7. A stream system flows through a breach in the rim into the crater from the higher ground to the south.

Crater 7 (the 'Main Crater')

The largest crater, Crater 7, is roughly elliptical, with dimensions of 180 m by 140 m. It apparently was formed by two adjacent projectiles. As in the case of the other larger craters at Henbury, the Main Crater shows both

HENBURY: A low-altitude aerial view of the southern craters at Henbury.

HENBURY: Henbury crater 7, the Main Crater, from its southeastern rim.

folding and thrust faulting, though they are more difficult to trace in the rims than at the Barringer Crater. A section of overturned rim flap is well-exposed along the south rim. A spur inside the south wall of the crater where the two craters intersected shows gentle folds of sandstone beds.

HENBURY: Henbury crater 6, the Water Crater, looking east towards the common rim of craters 6, 7, and 8.

Crater 8

Crater 8 is nestled close to the Water Crater and the Main Crater, with which it shares its north and west rims. It is 75 m in diameter and is relatively barren, with only two or three trees near its center. Its floor is of lighter albedo than those of the other large craters. The unshared rims rise to about 2 m from the exterior level. Ejecta clearly extends out to the east and south.

Crater 9

This 'crater' was considered doubtful by Alderman and cannot be identified now. There probably was not a crater at the indicated position, just east of Crater 8.

Crater 10

Crater 10 is one of the three smaller craters found on the ridge to the south of the main craters. It is isolated and well exposed and preserved. On the west side the beds of sandstone are tipped up and turned over at the rim and the east wall shows similar deformation, with one sandstone bed having been deformed into a broad anti-

HENBURY: Henbury crater 3 from its north rim, looking south.

HENBURY: Henbury crater 5 from its east rim, which is barely perceptible.

HENBURY: Henbury crater 4 from its floor near the south wall, looking towards the western rim.

HENBURY: Henbury crater 10 from the north. This crater was excavated in 1932.

HENBURY: Henbury crater 11 from the east.

HENBURY: Henbury crater 12, situated on a hillside, from its low, west rim.

cline and elsewhere on the rim there are alternating zones of twisted beds and fractures.

Crater 11
This crater is 15 m in diameter and was excavated in the 1930s in an unsuccessful search for large meteorite fragments.

Crater 12
Crater 12 lies on the slope of a sandstone ridge and has highly disparate rims on its north and south sides. The inner wall on the north rises 5 m above the floor, while that on the south is barely discernible. A stream bed lies just to the south of the south rim. The crater is 30 m in diameter. The rims show overturned strata, especially at the north. The floor is made up of smooth, white sediment and there are small trees and bushes along the walls.

Crater 13
This crater has very little rim. It is about 6 m in diameter and nearly a meter deep. It was excavated in the 1930s and a large meteorite was extracted from it.

Ejecta are conspicuous at the Henbury Craters, especially around the three largest craters. From aerial photographs, the ejecta blanket can be seen to extend out to distances averaging about 200 m from Crater 7. An impact melt called 'Henbury Glass' is found around the main craters, primarily on the north rim of Crater 7 and in a strip extending more than a kilometer east of the craters.

Access: The Henbury Craters are protected by the Northern Territory Conservation Commission as the Henbury Meteorite Craters Conservation Reserve. They are reached by a gravel road that turns off of the Stuart Highway 136 km south of Alice Springs. A right turn 13 km along this road takes the visitor 5 km to the north to a parking area, where there are minimal visitor facilities. The road is not recommended after a rain.

References

Alderman, A. R., 1932, The meteorite craters at Henbury, Central Australia. *Minerol. Mag., 23,* 19–32.

Hodge, P., 1965, The Henbury Meteorite Craters. *Smithsonian Contr. Astrophys., 8,* 199–213.

Hodge, P. and Wright, F., 1971, Meteoritic particles in the soil surrounding the Henbury Meteorite Craters. *J. Geophys. Res., 76,* 3880–3895.

Milton, D., 1965, Structure of a ray crater at Henbury Northern Territory, Australia. *Geol. Surv. Prof. Pap., 525-C,* 5–11.

Milton, D., 1968, Structural geology of the Henbury Meteorite Craters, Northern Territory, Australia. *Geol. Surv. Prof. Pap., 599-C,* 1–17.

Kelly West

Northern Territory

Lat/Long: S19° 56′, E133° 57′
Diameter: 10 km
Age: >550 Ma
Condition: Highly eroded

Kelly West is recognized as an impact structure through the presence of shatter cones in the rocks of the central peak, which is the only portion of the structure exposed.

Access: Kelly West is about 10 km south of Tennant Creek, near the Stuart Highway.

Reference

Tonkin, P. 1973, Discovery of shatter cones at Kelly West near Tennant Creek, Northern Territory, Australia. *J. Geol. Soc. Austr.*, *20*, 99–102.

Lawn Hill

Queensland

Lat/Long: S18° 40′, E138° 39′
Diameter: 20 km
Age: >540 Ma
Condition: Eroded

Adjacent to Lawn Hill National Park, the Lawn Hill impact structure was recognized in 1987. There is a ring of middle Cambrian limestones within the rim of the crater structure. The center has shatter cones and some evidence of melt rock.

Access: Lawn Hill is in northeast Queensland near the border with Northern Territories. There is a road to Lawn Hill from Mount Isa to the south and there is a small airport.

Reference

Stewart, A. and Mitchell, K. 1987, *Austr. J. Earth Sci.*, *34*, 477–485.

Liverpool

Northern Territory

Lat/Long: S12° 24′, E134° 03′
Diameter: 1.6 km
Age: 150 Ma
Condition: Partly exposed

The Liverpool Crater lies in a swampy area of Arnhem Land in the northern part of Northern Territory. It may be the only known meteorite crater to be infested with crocodiles, a condition that has made extensive field investigation a rare occurrence. It is on the flood plain of the Liverpool River and is surrounded by swamplands and meandering streams (the location of the Liverpool crocodiles).

The structure consists of a circular ring of sandstone breccia, 50 m high and 100–300 m wide. The breccia includes blocks of sandstone ranging in size from microscopic up to more than 5 meters across. Bedrock exposures of Proterozoic sandstone just exterior to the ring show strikes tangential to the ring and dips (inwards and outwards) at moderate angles. The interior is covered with post-crater deposits of soft sandstone. The breccia shows grains that have suffered shock deformation to various degrees.

Access: The Liverpool structure is not easily visited. In dry winters it is possible to reach it overland. It is on Aboriginal land and an entry permit is required. Darwin lies 330 km almost due west, and the Arnhem Highway leads from Darwin 250 km to the Border Store, beyond which travel is more difficult.

Reference

Guppy, D. J., Brett, R. and Milton, D. J., 1971, Liverpool and Strangways Craters, Northern Territory: two structures of probable impact origin. *J. Geophys. Res.*, *76*, 5387–5393.

Mt. Toondina

South Australia

Lat/Long: S27° 57′, E135° 22′
Diameter: 3 km
Age: <100 Ma
Condition: Eroded

Mt. Toondina was originally thought to be a salt diapere, but recent exploration of its central uplift and its gravity profile indicates that it has the form of an impact basin. The impact punched through lower Cretaceous shale. The central uplift is well-exposed and shows convergent flow, thrusting and puckering of Cretaceous and Jurassic beds, extending down to Precambrian.

Access: Mt. Toondina is about 150 km east of the Stuart Highway, northeast of Coober Pedy.

Reference

Shoemaker, E. M. and Shoemaker, C. S., 1987, Impact Structures of Australia. *Lunar Planet. Sci.*, *19*, 1079–1080.

Piccanniny

Western Australia

Lat/Long: S17° 26′, E128° 26′
Diameter: 7 km
Age: <360 Ma
Condition: Highly eroded

The Piccanniny structure is difficult to discern from the air or, casually, from the ground. There is no basin, no raised rim, and no obvious circular symmetry. The structure is clear only when geologically mapped. It lies on a plateau that is cut by deep ravines and that ends abruptly on the west in a series of cliffs.

PICCANNINY: The Piccanniny structure is located near the center of this aerial view of the plateau that is cut by the Piccinniny Gorge.

Access: The Piccanniny structure lies within the boundaries of Bungle Bungle National Park, which features unusual erosional features, most of which are south and east of the Piccanniny structure. The Spring Creek Track leads to the area from the Great Northern Highway. It is about 60 km on this rough track to the Piccanniny Gorge, which lies at the southwestern edge of the Piccanniny Structure.

Reference

Shoemaker, E. S. and Shoemaker, C. S., 1987, Impact structures of Australia. *Lunar Planet. Sci*, *19*, 1079–1080.

Snelling Crater

Western Australia

Lat/Long: S19° 24′, E127° 46′
Diameter: 0.03 km
Age: Young
Condition: Fresh

The Snelling Crater was identified by its structure through trenching in 1990, though its existence had been known to the local ranchers some years previously. The impacting meteorite hit gravel stream sediments and ejecta can be traced out to about 30 m from the center. No meteorite fragments have yet been reported and a preliminary soil survey did not turn up any microscopic meteoritic material.

SNELLING: The small feature tentatively identified as the Snelling Crater lies in these woods, near Wolfe Creek.

Access: The Snelling Crater is on the Caranya Station, about 15 km south of the Caranya homestead. It is easily accessible from Halls Creek (see Wolfe Creek Crater).

Reference

Shoemaker, E. S. and Shoemaker, C. S., 1990, personal communication.

Spider
Western Australia

Lat/Long: S16° 44′, E126° 05′
Diameter: 13 km
Age: >700 Ma
Condition: Eroded

The Spider structure gets its name from a remarkable feature in its center. Outcrops there have the shape of radiating 'legs' of rock. These are interpreted to be sections of sandstone that have been caused to slide over each other after the impact, which may have been at an oblique angle. Highly eroded walls encircle the central basin.

SPIDER: An aerial view of Spider. The structure extends across this photo and the remarkable folded central uplift is visible near the center of this view.

Access: Spider lies in a trackless area 240 km northwest of Halls Creek and can be reached from the Darby Gibb River Road, turning right near the Mt. Barnett homestead and proceeding overland.

Reference
Shoemaker, E. M. and Shoemaker, C. S., 1987, Impact structures of Australia. *Lunar Planet. Sci., 19*, 1079–1080.

Strangways
Northern Territory

Lat/Long: S15° 12′, E133° 35′
Diameter: 26 km
Age: >540 Ma
Condition: Highly eroded

The Strangways structure is a circularly-symmetric formation, difficult to discern either from the ground or from the air because of its eroded and largely sediment-covered condition. There is a ring pattern of tangentially striking beds of sandstone, with high ridges alternating with flat valleys. Shatter-coning is found in some of the sandstone outcrops. The center of the structure is largely covered by post-crater sediments, but some outcrops of breccia are present, including a 1 km by 2 km hill that is 50 m high. Quartz grains commonly show shock features, including planar features, cleavage, and recrystalization. The chemistry of the matrix of the breccia shows an anomalously high nickel content, which may be the result of meteoritic contamination. An outcrop of gneiss near the center suggests considerable uplift, as this basement rock is buried deeply by sediments in the surroundings.

Access: Strangways is difficult to visit. It lies just west of the Strangways River about 75 km southeast of the Elsey Station homestead, which is 35 km east of Mataranka off the Roper Highway.

Reference
Guppy, D. J., Brett, R. and Milton, D. J., 1971, Liverpool and Strangways Craters, Northern Territory: two structures of probable impact origin. *J. Geophys. Res., 76*, 5387–5393.

Teague Ring
Western Australia

Lat/Long: S25° 52′, E120° 53′
Diameter: 28 km
Age: 1685 Ma
Condition: Eroded, partly exposed

The Teague Ring is a large impact feature in the wilderness desert of Western Australia. It is highly eroded, with

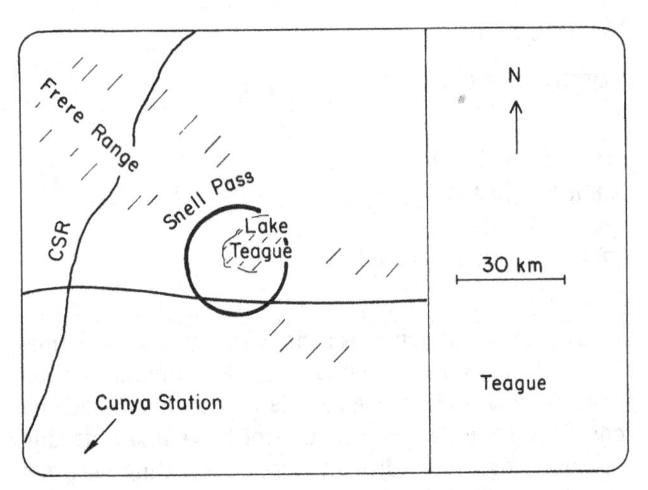

TEAGUE RING: Location map for the Teague Ring impact structure. CSR marks the location of the historical Canning Stock Route, blazed in the 19th century, but used now only by hardy adventurers.

TEAGUE RING: An outcrop of syenite at the central uplift of the Teague Ring, showing a ridged surface that is similar to shatter-coning.

only a few outcrops rising above the sand dunes at the center. The structure bisects a shallow mountain range that extends northwest to southeast, in which it forms a circular depression.

Evidence that this is an impact crater can be found in the outcrops in the depression and along its edge. In the outer parts of the structure there are outcrops of country rock that show tilted beds, sloped away from the center. Outcrops of pink granite and coarse and fine-grained syenite from within the depression show the effects of shock metamorphism.

The crater floor is largely sand, with outcrops of granite, syenite and younger sediments. There is a 10-km-wide central uplift of syenite, which shows near-shatter-coning and pseudodacylite. There are scattered clumps of trees and lots of kangaroos in the depression.

Access: Teague can be reached overland on rough desert tracks from the town of Wiluna, which is about 164 km to the southwest, via the Cunyu and Mt. Alice Stations.

References

Butler, H., 1974, The Lake Teague ring structures, Western Australia: an astrobleme? *Search, 5*, 534–536.

Shoemaker, E. and Shoemaker, C., 1988, Impact structures of Australia. *Lunar Planet. Sci., 19*, 1079–1080.

Veevers Crater

Western Australia

Lat/Long: S22° 58′, E125° 22′
Diameter: 0.08 km
Age: <1 Ma
Condition: Relatively fresh

Discovered by prospectors who were flying over the area in 1981, the Veevers crater is a well-preserved small crater in the Gibson Desert. It has a slightly raised rim, about 4 m above the surroundings, and a flat, sediment-filled crater floor about 10 m below the rim. The meteorite that caused the impact was a medium octahedrite. About 35 fragments of it were recovered by Caroline Shoemaker, mostly from an area off the northeast rim. The country rock is mid-Tertiary laterite, which, being iron-rich, makes the search for meteorite fragments difficult. The area is frequented by wild camels.

Access: Veevers is about 1000 km northeast of Wiluna via desert roads. On the rough desert track that goes north

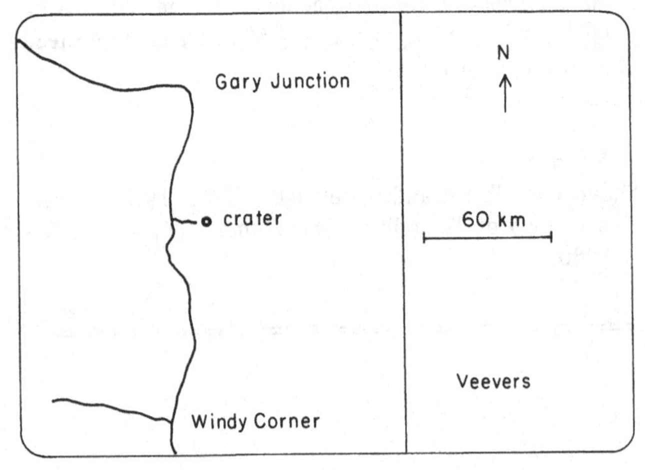

VEEVERS: Location map for the Veevers crater.

VEEVERS: The floor of the Veevers crater from the west rim.

VEEVERS: A meteoriticist writes notes at the south rim of the Veevers crater.

from Everard Junction, one proceeds 80 km past Windy Corner (about 30 km past the Tropic of Capricorn sign) and then turns east on a narrow track, which is followed 15 km to the crater. At least one high sand dune must be crossed *en route*.

Reference

Shoemaker, E. and Shoemaker, C., 1988, Impact structures of Australia. *Lunar Planet. Sci.*, *19*, 1079–1080.

Wolfe Creek

Western Australia

Lat/Long: S19° 18′, E127° 47′
Diameter: 0.88 km
Age: <0.3 Ma
Condition: Well preserved

The Wolfe Creek Crater (usually misspelled as 'Wolf Creek') is Australia's most famous impact crater. Nearly the size of the Barringer Crater, it is young enough to have suffered only mild erosion, retaining most of its structure. It is a popular tourist attraction, being reasonably accessible from the town of Hall's Creek. Several tourists a day may visit it at the peak of the season (e.g., in early spring). Although it lies within the boundaries of an immense ranch, the Carranya Station, it is set aside as a National Park. Furthermore, to add to its fame, it is the

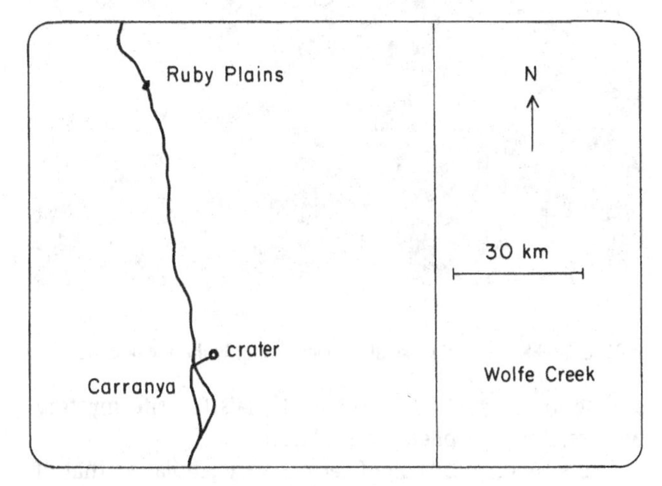

WOLFE CREEK: Location map for the Wolfe Creek crater.

WOLFE CREEK: Aerial view of Wolfe Creek crater from the west.

WOLFE CREEK: Wolfe Creek crater, looking east from the west rim.

WOLFE CREEK: Uptilted rocks at the west rim of Wolfe Creek crater.

setting for a novel by one of Australia's favorite mystery writers, Arthur Upfield (see references).

The structure of Wolfe Creek is very similar to that of Barringer. The rim rises 35 m above the surrounding plain, forming a perfect ring of raised beds. The rocks at the top of the rim are vertical and overturned beds of the target rocks: Devonian sandstone and laterite. Plastic deformation is evident. Meteoritic shale balls can be seen on the rim, some welded to the laterite layer. The parent meteorite was an octahedrite.

To the south and southeast, sand dunes nearly reach the top of the rim. In a few places they have spilled over and flowed to the crater floor, which lies 55 m below the rim and is flat and tree-covered (it retains rainwater well). The true crater floor is about 150 m below the present surface.

The following is a trail log for the boot-beaten path along the top of the rim. As opposed to the Barringer Crater trail, (Chapter 2), this one is taken counter-clockwise (it is in the southern hemisphere!).

Distance from the parking area (km)	Features to note
0.0	The trail passes outcrops of the laterite that covers most of the Wolfe Creek area.
0.2	The western rim affords a spectacular view of the crater. Proceed to the right.
0.3	To the right on the outer slopes at this point examples can be found of shale balls welded into laterite.
0.4	Directly across the crater, the far inner wall shows bedding of gently dipping quartzite.
0.6	The inner walls beneath you at this point are steeply dipping Devonian sandstone. The erosion rate has been measured at only 2 cm per 1000 years. Along this part of the rim it is possible to see completely overturned rocks, with the bedding layers having suffered a nearly 180° flip.
0.8	Nearing the south rim, larger sheets of uptilted sandstone are encountered. A few shale balls can be seen near here, just outside the rim crest.
0.9	Vertical beds of sandstone show twisted, distorted shapes. The laterite layer is upside-down, with occasional shale balls imbedded, though they are rare.
1.1	Ejecta extends out to the south from the rim, which is almost breached by the sand dunes in places. Most of the rather rare unoxidized meteorite fragments were found south and west of here. The impacting object was an iron meteorite, classed as a medium octahedrite.
1.6	The entire east rim is encroached by the extensive sand dunes, which have had their westward progress inhibited by the raised rim of the crater.
2.0	Along the northeast part of the rim, beds of sandstone are standing upright, tilted somewhat more than 90° from their original nearly flat attitude.
2.3	The north rim, like the west, has a more gentle inner slope than the east. The meteorite came from an easterly direction.
3.0	Return to the west rim and the trail down to the road.

WOLFE CREEK: Wolfe Creek crater from the south rim, looking east.

WOLFE CREEK: A ball of meteorite shale located *in situ* on the south rim of Wolfe Creek crater.

Access: Wolfe Creek Crater can be reached by an unpaved road that leaves the Great Northern Highway about 20 km west of Halls Creek. Proceed south about 100 km on this road to a short spur road to the east, near the Carranya Station.

References

Cassidy, W. A., 1954, The Wolfe Creek, Western Australia, Meteorite Crater. *Meteoritics, 1*, 197–199.

Guppy, J. D. and Matheson, R. S., 1950, Wolfe Creek, Western Australia., *J. Geol., 58*, 30–36.

McNamara, K., 1982, *Wolfe Creek Crater.*, Western Australian Museum, Perth, 16 pp.

Reeves, F. and Chalmers, R. O., 1949, The Wolfe Creek Crater, *Austr. J. Sci.* 11, 154–156.

Upfield, A., 1934, *Death of a Swagman.* Scribners, New York, 334 pp.

6

Impact structures of Europe

Azuara

Spain

Lat/Long: N41° 10′, W00° 55′
Diameter: 30 km
Age: <130 Ma
Condition: Eroded

The Azuara structure was recognized as an anomalous feature in the Spanish landscape in 1983, when two hypotheses were put forward: that it is an impact structure and that it is a tectonically isolated block of endogenic origin. The issue was settled by the discovery of impact breccia that contains highly-shocked quartz grains.

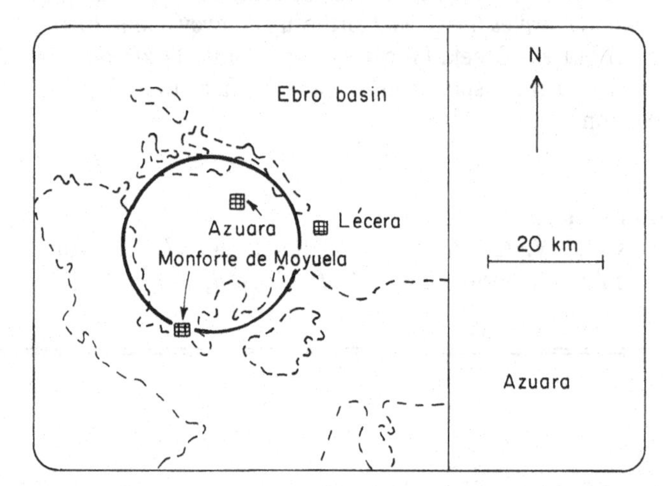

AZUARA: Sketch map for the Azuara structure (after Ernston *et al.*).

The Azuara structure consists of a partial ring of Mesozoic folded rocks, characterizing the area to the southwest, enclosing more recent sediments, which characterize the country to the northeast. The rim shows inverted stratigraphy and strata that dip toward the center. Outcrops of highly deformed breccias are found, for example, among the Ordovician quartzites and the Jurassic limestones. Microscopic examination has shown the presence of abundant planar features in quartz grains in the breccia, with multiple orientations.

Access: The structure is roughly centered on the village of Azuara, which is about 50 km south of the city of Zaragoza in northeast Spain. The outcrops of the impact-distorted precrater rocks are primarily in the southwest portion of the structure, near the villages of Monteforte de Moyuela, Nogueras and the area just west of Herrera de los Navarros.

Reference

Ernstson, K., Hammann, W. Fiebag, J. and Graup, G., 1985, Evidence of an impact origin for the Azuara Structure (Spain). *Earth Plane. Sci. Lett.*, 74, 361–370.

Boltysh

Ukraine

Lat/Long: N48° 45′, E32° 10′
Diameter: 25 km
Age: 88 Ma
Condition: Slightly eroded and entirely buried

The Boltysh Crater is located in the Precambrian metamorphic rocks of the Ukrainian Shield. It is covered with more than 500 m of post-cratering sediments. There is a 6 km diameter central uplift that extends about 500 m above the encircling trough. Little erosion seems to have occurred in the millennia after the crater formed, as there is a fairly complete set of impact rocks (breccia and impact melt) directly below the post-crater sedimentary rocks. The impact melt sheet is 200 m thick and shows three horizontal zones, a central glassy zone with a microcrystalline zone below and above it. This arrangement is the opposite of the normal situation, in which microcrystalline rocks are sandwiched between zones of glass in melt rock layers. There is an enhancement of nickel, chromium and iridium in the melt rocks with respect to the target rocks. Coesite has been discovered in samples of breccia, both in those underlying the impactites and in samples of ejecta collected from the Tyasmin River basin.

Access: The Boltysh structure is near Kirovograd, about 200 km south of Kiev.

References

Grieve, R. A. F., Reny, G., Gurov, E. P. and Ryabenko, V. A., 1987, The melt rocks of the Boltysh Impact Crater, Ukrain, USSR., *Contrib. Mineral. Petrol.*, *96*, 56–62.

Gurov, Y. P., Valter, A. A. and Rakiskaya, R., 1980, Coesite in rocks of meteorite explosion craters on the Ukrainian Shield, *Int. Geol. Rev.*, *22*, 329–331.

Dellen

Sweden

Lat/Long: N61° 48′, E16° 48′
Diameter: 15 km
Age: 110 Ma
Condition: Eroded, lake-covered

A well-studied crater, the Dellen structure has been investigated by gravity, magnetic, and electromagnetic surveys. It shows the presence of coherent impact melt bodies near the center of the structure. Presently most of the crater basin is occupied by two lakes, the Norra Dellen and the Sodra Dellen. An age of 110 Ma was determined from the melt rocks using the $^{40}Ar–^{39}Ar$ method, while a Rb–Sr age of 90 Ma has been reported.

Access: The Dellen lakes are easy to visit. They are in central Sweden near the coast of the Gulf of Bothnia, about 25 km west of the coastal city of Hudiksvall.

Reference

Muller, N., Hartung, J., Jessberger, E. and Reimold, W., 1990, $^{40}Ar–^{39}Ar$ ages of Dellen, Janisjarvi and Saaksjarvi impact craters. *Meteoritics*, *25*, 1–10.

Dobele

Latvia

Lat/Long: N56° 35′, E23° 15′
Diameter: 4.5 km
Age: 300 Ma

The Dobele structure is near the town of Dobele, about 75 km south of Riga.

Reference

Grieve, R. A. F., 1991, Terrestrial impact: the record in the rocks. *Meteoritics*, *26*, 175–194.

Gardnos

Norway

Lat/Long: N60° 39′, E09° 00′
Diameter: 5 km
Age: ~400 Ma
Condition: Highly eroded and partially covered

The Gardnos structure was recognized as of probable impact origin in 1991. Previously, it was thought that the peculiar brecciation found there had an explosive volcanic origin. The structure is a bowl-shaped semi-circular hanging valley. Although largely covered with recent glacier sediments and forest growth, the rocks of the structure are exposed in many small stream valleys, which reveal both the many post-crater rock beds and the deformed crystalline rocks of the crater's skeleton. There are impact breccias and melt rocks that are similar to suevite. Quartz and feldspar grains reveal many planar features and other shock metamorphism is evident.

Access: The structure is centered 9 km north of the village of Nesbyen in south-central Norway. It is about 120 km northwest of Oslo in the Hallingdal valley.

Reference

Dons, J. and Naterstad, J, 1992, The Gardnos impact structure, Norway. *Meteoritics, 27,* 215.

Gusev

Russia

Lat/Long: N48° 21′, E40°, 14′
Diameter: 3.5 km
Age: 65 Ma
Condition: Eroded

The Gusev structure has the form of a small ellipsoidal basin. It is filled with megabreccias to a depth of 400 m. It lies only about 1 km to the north of the edge of the Kamensk structure. Because of its proximity and its similar age, it is assumed that Gusev and Kamensk formed at the same time, possibly by collision of the Earth with a double asteroid, or an elongated asteroid that broke in two in or above the atmosphere.

Access: The structure lies near the town of Kamensk-Shakhtinskiy, just east of the border between Ukraine and Russia.

Reference

Masaitis, V. L., 1976, Astroblemes in the USSR. *Int. Geol. Rev., 18,* 1249–1258.

Ilumetsy

Estonia

Lat/Long: N57° 58′, E27° 25′
Diameter: 0.08 km
Age: >0.002 Ma
Condition: relatively fresh

There are three recognized craters at Ilumetsy (or Ilumetsa, as it is currently called). Two of these are fresh and well-exposed; both were first recognized and explored in 1938. The third was a more recent discovery and it is flat. The country rock is glacial till and Devonian sandstone. Two of the craters are filled with peat.

The individual craters are named and described as follows:

Porguhaud
This, the largest crater, is 80 m in diameter and is filled with peat. There is a lenticular layer of shattered Devonian bedrock mixed with Quaternian sand and clay. The rim shows uplifted beds and ejecta. The rim height is 4 m and the center of the crater is 12 m deep.

Sugavhaud
This crater has a diameter of 50 m, a depth of 4 m and a rim height of between 1 and 2 m. It is not peat-filled.

Tondihaud
Discovered in 1957, this filled-in crater is 24 m in diameter. It has no rim and the center has a 1.5 m-deep fill of peat.

Access: The Ilumetsy craters are near the southeast border of Estonia, about 175 km southeast of Tallinn, west of Lake Pskov. The area is marshy. There are discrepancies about the latitude and longitude in the literature; the position given here agrees with the designation 'Ilumetsa' shown on the 1:250 000 topographic map.

Reference

Pirrus, E. and Tirrmaa, R., 1990, The meteorite craters in Estonian, in *Abstracts, Fennoscandian Impact Structures*, vol. 51, ed. L. Pesonen and H. Niemisara. Geological Survey of Finland.

Ilyinets

Ukraine

Lat/Long: N49° 06′, E29° 12′
Diameter: 4.5 km
Age: 395 Ma
Condition: Highly eroded and partially covered.

Most of Ilyinets is gone. Only a thin layer of suevites and breccias exists, mostly covered by a layer of sands and Devonian clays. The breccias consist of fragments of Precambrian rocks of the Ukrainian shield and they are underlain by fractured granites. Shock-metamorphosed mineral grains are contained in the fragments of crystalline rocks recovered from the breccias.

It has been suggested that Ilyinets may be only one (the largest) of a group of craters, as rocks similar to the Ilyinets impactites have been recovered from six small areas to the south and southwest of the crater.

Access: Ilyinets is in central Ukraine, about 200 km south of Kiev and 45 km southeast of the city of Vinnitsa.

Reference

Masaitis, V. L., 1976, Astroblemes in the USSR. *Int. Geol. Rev., 18*, 1249–1258.

Janisjarvi

Russia

Lat/Long: N61° 58′, E30° 55′
Diameter: 14 km
Age: 700 Ma
Condition: Eroded, lake-filled

Lake Janisjarvi lies near the present Finnish border of Russia on the Baltic Shield. Its islands show evidence of an impact origin, including the presence of massive impact glass and impact breccias. About 20% of the crystallized glass includes partially remelted fragments of the Proterozoic schists that make up the target rock. Shatter cones have been recovered and quartz grains show planar features indicating shock metamorphism.

Access: Lake Janisjarvi (sometimes labeled Ozero Yanisyarvi) is in Karelia, 225 km north of St. Petersberg and just north of Lake Ladoga. A rail line runs near its south shoreline.

References

Masaitis, V., Sindeyev, A. and Staritskii, Y., 1977, Impactites from the Yanisyarvi astrobleme. *Meteoritics, 12*, 84.

Muller, N., Hartung, J., Jessberger, E. and Reimold, W., 1990, ^{40}Ar–^{39}Ar ages of Dellen, Janisjarvi and Saaksjarvi impact craters. *Meteoritics, 25*, 1–10.

Kaalijarvi Craters

Estonia

Lat/Long: N58° 24′, E22° 40′
Diameter: 0.110 km
Age: 0.004 Ma
Condition: Fresh, slightly altered

The Kaali craters include the largest, which is lake-filled (thus the name Kaalijarvi), and eight smaller craters. The

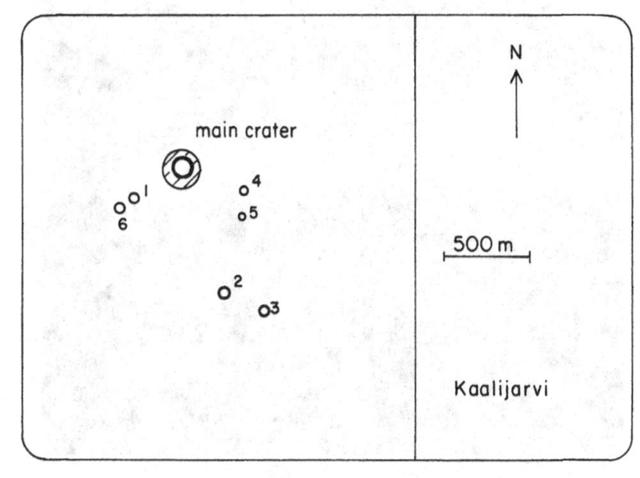

KAALIJARVI: Map of the locations of the craters at Kaali. Kaalijarvi is the lake in the main crater.

KAALIJARVI: Low-altitude aerial view of the main crater at Kaali. Farm buildings are adjacent to the lake-filled main crater (photo courtesy of R. Tiirmaa).

KAALIJARVI: The main crater at Kaali (photo courtesy of R. Tiirmaa).

KAALIJARVI: The lake in the center of the main crater, called Kaalijarvi (photo courtesy of R. Tiirmaa).

large crater is 110 m in diameter and about 20 m deep, while the smaller craters are shallow with diameters that range from 12 to 40 m. Meteorite fragments found in the smaller craters range in size from microscopic up to 20 g and they indicate that the incoming body was a coarse octahedrite. The soil surrounding the craters contains many small (<1 mm) fragments and spherules of meteoritic and magnetic silicate material, with an indication of a higher density of these objects along the inferred incoming trajectory (from the east–northeast) than elsewhere. The age of the craters has been estimated from the presence of small mollusk shells, which indicate that they were formed soon after the retreat of the post-glacial sea, about 4000 years ago.

The individual craters have been described as follows:

Main Crater

Discovered in 1827, the Main Crater is 110 m in diameter and has a rim about 5 m high. It is filled with a lake, the floor of which is a thick layer of mud and peat. The crater is surrounded by a ring of trees and bushes, making it stand out in the fields like an oasis in the desert. Dolomite beds are upturned around the rim at an angle of about 60°. Drilling in 1929 showed that the dolomite beneath the center of the crater is crushed to a depth of about 5 m and is undisturbed beneath that.

Crater 1

Discovered in 1927, the largest secondary crater is 39 m in diameter, 4 m deep with no rim.

Craters 2/8

With diameters of 36 and 25 m, respectively, these twins were discovered in 1927. They are 3 m deep with no raised rim. Crater 2 was partially excavated in 1937 and it was found to have first a 10-cm layer of soft soil,

KAALIJARVI: Dolomite layers in the rim of the main Kaali crater, showing disturbance caused by the impact (photo courtesy of R. Tiirmaa).

underlain by a mix of rock flour, dolomite fragments and glacial deposits. Fragments of iron meteorite (medium octahedrite) were encountered at a depth of 1 m. A total of 28 meteorite fragments, weighing 102 gm in total, were eventually extracted from Crater 2.

Crater 3
With a diameter of 33 m and a depth of 3 m, this crater was discovered in 1927. A magnetic survey made in 1955 showed only very small anomalies, probably caused by small meteorite fragments at shallow depths.

Crater 4
Twenty meters in diameter and 1 m deep, this crater was found to have meteoritic material in it. It was discovered in 1927.

Crater 5
Three small iron meteorites were recovered in the initial excavations of this crater. It is 13 m in diameter and 1 m deep.

Crater 6
This shallow, rimless crater, 26 m in diameter, was discovered in 1937.

Crater 7
The most-recently-recognized crater, this 1-m-deep, rimless crater was found in 1965.

Access: The craters are on the Island of Saarimaa, 25 km northeast of the coastal city of Kuressaare (also known as Kingisepp), and are adjacent to a farm road. They are in a geological preserve.

KAALIJARVI: Kaali crater number 3 (photo courtesy of R. Tiirmaa).

Reference

Tiirmaa, R., 1992, Kaali craters of Estonia and their meteoritic material. *Meteoritics*, *27*, 297.

Kaluga

Russia

Lat/Long: N54° 30′, E36° 15′
Diameter: 15 km
Age: 380 Ma
Condition: Buried

The Kaluga structure is buried beneath 800 m of sedimentary rocks (mostly Devonian and Carboniferous strata). It has a bowl shape and is surrounded by deformed older Devonian sedimentary rocks. There is a layer of suevites and breccias about 100 m thick in the crater basin. There is an asymmetrical uplift on one side, rising about 300 m above the bottom of the basin and made up of breccias derived from the sedimentary and crystalline country rock. Kaluga is one of the most deeply-buried impact structures yet known.

Access: Although there is little to see of the structure at the surface, its location is relatively easy to visit, as it lies beneath the city of Kaluga, about 160 km southwest of Moscow.

Reference

Masaitis, V. L., 1976, Astroblemes in the USSR, *Int. Geol. Rev.*, *18*, 1249–1258.

Kamensk

Russia

Lat/Long: N48° 20′, E40° 15′
Diameter: 25 km
Age: 65 Ma
Condition: Eroded and buried

The Kamensk structure lies adjacent to the smaller Gusev crater. The country rock includes folded limestones, shales, and marls of Permian and Lower Triassic age. The crater is overlain with more than 250 m of Paleocene deposits. There is a central uplift that rises about 400 m above the surrounding trough. The whole depression is filled with a layer of impact breccia, including blocks as large as 10 m.

Access: The structure lies under the surface near the city of Kamensk-Shakhtinskiy, near the Ukraine–Russian border.

Reference

Masaitis, V. L., 1976, Astroblemes in the USSR. *Int. Geol. Rev., 18*, 1249–1258.

Karla

Russia

Lat/Long: N54° 54′, E48° 00′
Diameter: 12 km
Age: 10 Ma
Condition: Eroded, partially filled

The Karla structure is found in horizontal beds of sedimentary rocks from the Mesozoic. The crater is defined by a ring-shaped depression that is filled by about 500 m

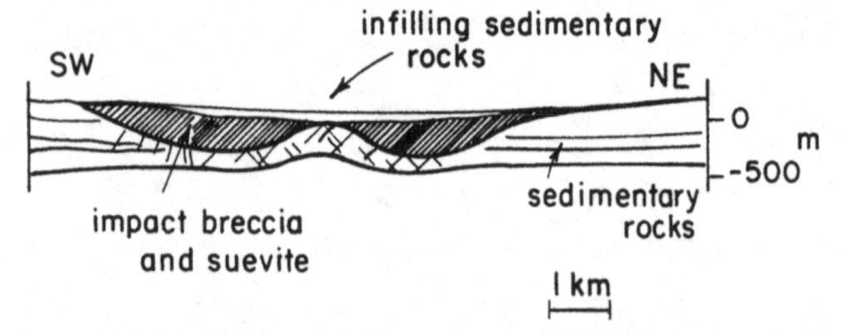

KARLA: Profile of the Karla structure (after Masaitis *et al.*)

of impact breccia. There is a central uplift about 3 km in diameter, which brings shattered middle Carboniferous limestone close to the surface. Expelled breccia surrounds the basin and extends about 9 km to the northeast. Shatter cones have been retrieved from exposures along the Karla River.

Access: The Karla structure is located in the valley of the Sviyaga River, a branch of the Volga. It is about 650 km east of Moscow.

References

Masaitis, V. L., Danilin, A. N., Karpov, G. M. and Raykhlin, A. I., 1976, Karla, Obolon, and Rotmistrovka astroblemes in the European part of the USSR. *Doklady Earth Sci., 230*, 48–51.

Kjardla

Estonia

Lat/Long: N59° 00′, E22° 42′
Diameter: 4 km
Age: 455 Ma
Condition: Buried

The Kjardla (or Kardla) Structure is on the Island of Hiiumaa, just to the north of the island that hosts the much more recent Kaalijarvi Craters. About 120 bore holes have been drilled and they show the characteristic features of a simple crater of this size. There is a layer of impact breccia in the crater proper, a small central uplift (50 m high), a raised rim and an external layer of outfall megabreccia. The breccia contains grains of quartz that show two to three systems of shock-induced planar systems. Impact debris can be traced out to 50 km from the crater. It is possible locally to buy bottled water that is derived from below the bottom of the crater and which is said to be exceptionally good.

Apparently the impact occurred in a shallow shelf of the Baltic epicontinental sea. The impact bored into limestone, sand and clay, and Precambrian crystalline rocks. After the impact, marine sediments were laid down on top of it.

Access: The basin lies below the area of the coastal village of Kjardla on the Island of Hiiumaa off the western coast of Estonia, about 100 km southwest of Tallinn. The center of the structure is about 3 km west of Kjardla.

Reference

Puura, V. and Suuroja, K., 1990, Ordovician impact crater at Kardla, Island of Hiimaa, Estonia, in *Abstracts, Fennoscandian Impact Structures*, ed. L. Pesonen and H. Niemisara. Geological Survey of Finland, 30 pp.

Kursk

Russia

Lat/Long: N51° 40′, E36° 00′
Diameter: 5.5 km
Age: 250 Ma
Condition: Buried

The Kursk structure is a circular basin that lies beneath about 200 m of Jurassic and Cretaceous strata. It has a central uplift that consists of crushed rocks of the Devonian and Carboniferous sequences that lie below the crater.

Access: The Kursk structure lies near the city of Kursk, which is about 450 km south of Moscow.

Reference

Masaitis, V. L., 1976, Astroblemes in the USSR. *Int. Geol. Rev., 18*, 1249–1258.

Lappajarvi

Finland

Lat/Long: N63° 12′, E23° 42′
Diameter: 17 km
Age: 77 Ma
Condition: Eroded, partially exposed

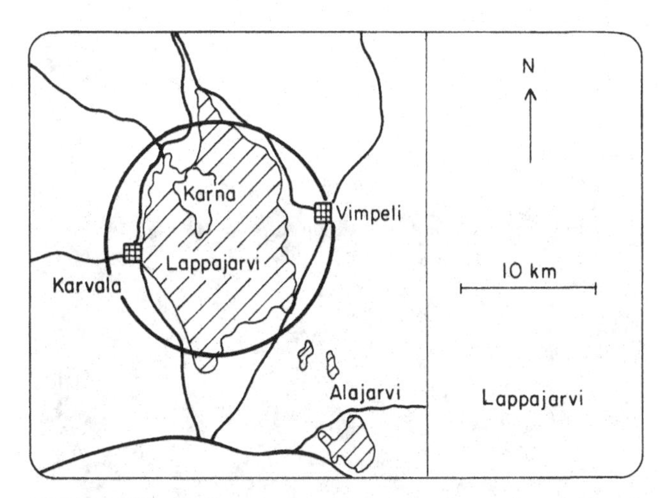

LAPPAJARVI: Map of Lappajarvi and the surrounding roads and villages.

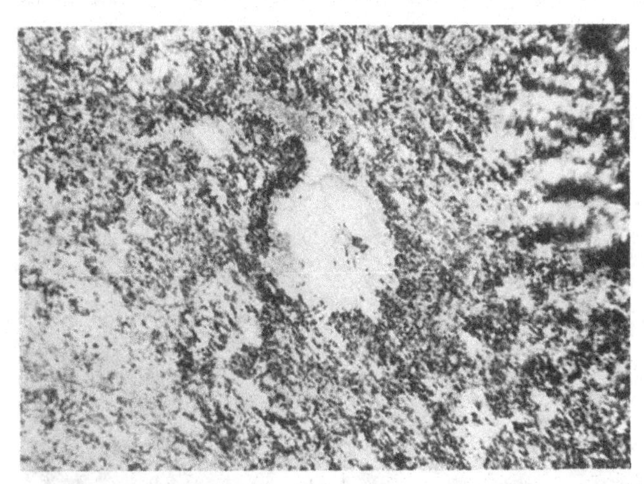

LAPPAJARVI: View of Lappajarvi from space (image from ERTS, an earth resources satellite, courtesy of NASA).

When coesite was found in the Ries suevite and geologists began to accept the impact hypothesis for that famous puzzle, other geological anomalies in Europe were re-examined. Following a geological congress in 1967, in Germany, where the new hypothesis about the Ries was discussed, the Finnish geologist Sahama returned to Helsinki wondering whether one of Finland's geological mysteries, the Lappajarvi structure, might have a similar explanation. Many years before, some Swedish circular features, such as Lake Mien, had been proposed as possible impact structures by N. B. Svensson, whose ideas, however, were not taken very seriously by his fellow geologists. But the new types of evidence, shock induced minerals, provided a way to settle the issue.

Lappajarvi ('Lappa Lake' in Finnish) is a 17-km-diameter lake in central Finland. Surrounded by a rim of forested hills, it nearly fills a flat circular plain. Around the lake are fertile farms, growing wheat and rye and potatoes. Many farms have typical Finnish farm houses with red wooden siding and white trim. In the groves of birch and spruce along the shore, summer cabins are hidden here and there. Some of these, on the shores of the central island of Karna, have outcrops of interesting rock of an unusual sort, called Karnaite, after the island's name. The rock is dark in color and seems to be made up of pieces

LAPPAJARVI: The lake of Lappajarvi, from the eastern rim of the structure.

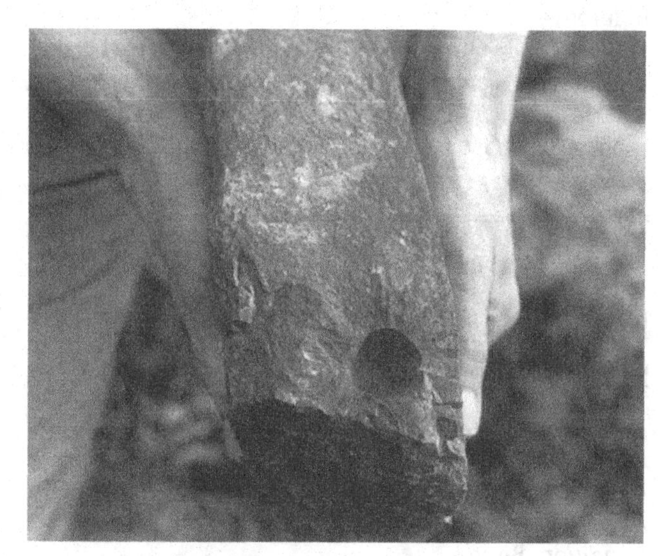

LAPPAJARVI: Karnaite, the impact melt rock found on the Island of Karna in Lappajarvi, sometimes shows the effect of bubbles.

LAPPAJARVI: Finish geologist M. Lehtinen pionts out clasts in impact breccia from Lappajarvi.

of once-shattered rocks and melts in a grainy matrix. Early geologists had trouble classifying karnaite; it was clearly like a breccia, but it had features indicating an episode of melting.

Southeast of the lake shore, near the village of Yakahari, the rim of the structure is high and steep enough that there is a ski resort there, complete with ski lift. Geologically, the rim is primarily made of uplifted (and often shattered) target rock, mostly granite, gneiss and schist. Pieces of quartzite can be found that have a laminar structure, caused by the explosive pressure of the impact. The roads that climb up to the southeast are mounting the hard skeleton of the crater rim. The top of the rim was scraped away long ago by the erosion of water and ice, but the rim relief is there around much of the circle. It is broken in the south, where ice-age glaciation has pushed the rock south, and broken in the north, where the outlet river now flows, eventually emptying into the Sea of Bothnia.

Sharp-eyed geologists have found rocks from the crater far to the south, with many pieces of karnaite turning up 40 or 50 km south of where the glaciers plucked them from the crater wall. There have even been reports of a few pieces of karnaite being found near Helsinki, 300 km to the south.

The karnaite fills in most of the crater structure. A glimpse into the structure below the surface was provided in 1988 by a drill hole that extended 217 m down below the surface of Karna Island. Geologists extracted cores from this drill hole and found a sequence of rock types that is typical for impact structures. Down to depths of 24 m the rock is upper karnaite. From there to 118 m it is perlitic karnaite. This is underlain by a 26-m-thick layer of 'lower karnaite' and under this is about 5 m of suevite, very like the glassy impact melt first found at the Ries crater. Below this is breccia, the top part finer than the bottom part, which is quite coarse.

Lappajarvi is a fairly old crater, dated by isotope analysis to have been formed 77 million years ago.

The Finnish geologist Marti Lehtinen spent his summers in the crater as a child, as his grandparents lived in the charming town of Vimpeli, just inside the eastern rim. It is a pleasant fact that he later did his doctoral thesis on the Lappajarvi structure and put together much of the geological evidence that we have of its true nature. He and Frederick Pipping of the Finnish Geological Survey have continued their work on this important example of the impact scars of the Baltic shield.

As in other impact structures, the Lappajarvi crater shows a strong gravity anomaly. A circular, negative anomaly is centered on it, with a 17 km diameter and a total mass deficit of 4×10^{16} g. This deficit arises because the Lappajarvi impact melt rocks and breccias are considerably less dense than the original bedrock. The mean density of the surrounding bedrock is 2.7 g/cm^3, while the karnaite, suevite, and impact breccia have densities between 2.3 and 2.5 g/cm^3.

Access: Lake Lappajarvi is easily visited. It is in central Finland, 100 km east of the coastal city of Vaasa. The village of Lappajarvi is on the northern edge of the lake and Vimpeli is on the eastern shore. There is at least one elegant hotel at the lake.

References

Lehtinen, M., 1976, Lake Lappajarvi: a meteorite impact site in Western Finland. *Geo. Surv. Finland, Bull., 282*, 92 pp.

Reimold, W., 1982, The Lappajarvi meteorite crater, Finland: petrography, Rb Sr, major and trace element geochemistry of the impact melt and basement rocks. *Geochim. Cosmochim. Acta, 46*, 1203–1225.

Svensson, N-B. 1971, Lappajarvi structure, Finland: morphology of an eroded impact structure. *J. Geophys. Res, 76*, 5382–5386.

Logoisk

Belarus

Lat/Long: N54° 12′, E27° 48′
Diameter: 17 km
Age: 40 Ma

The Logoisk structure is about 50 km north of Minsk.

Reference

Grieve, R. A. F., 1991, Terrestrial impact: the record in the rocks. *Meteoritics, 26*, 175–194.

Mien

Sweden

Lat/Long: N56° 24′, E14° 54′
Diameter: 9 km
Age: 121 Ma
Condition: Eroded and partly filled by Lake Mien

Lake Mien has long been known to have present a strange kind of rhyolite-like rock called 'Mien-rhyolit', which occurs commonly on an island in the lake named Ramso. Somewhat amazingly, although until recently most geologists believed this rock to be volcanic, A. G. Hogbom suggested as long ago as 1910 that it might have an impact origin. The question was resolved in 1965 when coesite was identified.

The Mien-rhyolit is an impact melt rock, which has interesting features, including many scattered vesicles. Some of these are filled by iron-rich spherules that range in size from submillimeter up to 3 cm. These features show evidence for a great deal of weathering and chemical alteration. The chemistry of the impact melt has led some investigators to suggest that the impacting body was a comet, rather than an asteroidal body.

Access: Lake Mien is easily accessible. It is in southern Sweden, about 20 km north of the coastal city of Karlshamn.

References

Ekelund, A. and Engstrom, E., 1990, Detection of cm sized spheroidal Fe–Mn bodies in rhyolite from Lake Mien, an impact site in Southern Sweden. *Lunar Planet. Sci., 19*, 297–298.

Svensson, M. and Wickman, F., 1965, Coesite from Lake Mien, Southern Sweden. *Nature, 205*, 1202–1203.

Misarai

Lithuania

Lat/Long: N54° 00′, E23° 54′
Diameter: 5 km
Age: 395 Ma

The Misarai structure is in southern Lithuania, near the Polish border.

Reference
Grieve, R. A. F., 1991, Terrestrial impact: the record in the rocks. *Meteoritics, 26,* 175–194.

Mishina Gora

Russia

Lat/Long: N58° 40′, E28° 00′
Diameter: 4 km
Age: <360 Ma
Condition: Eroded and buried

Mishina Gora (also called Mishinogorsk) lies in Paleozoic sedimentary rocks. A basin, about 700 m deep, is filled with impact breccias and fusion glasses. There are small shatter cones in some of the pieces of gneiss and schist located in the breccias. Quartz crystals and plagioclase glass show intense deformation.

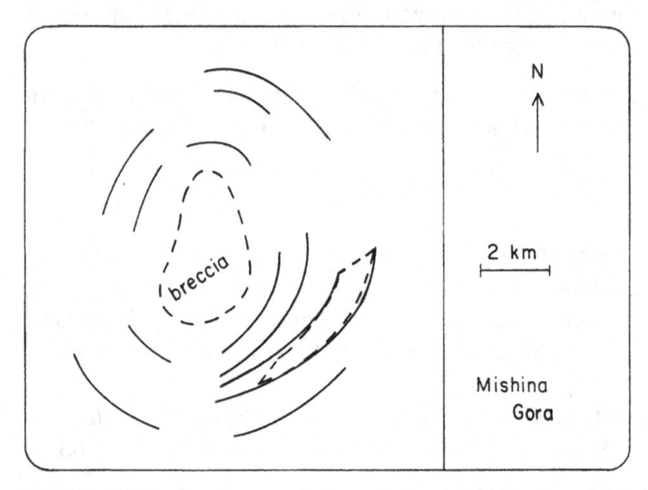

MISHINA GORA: Sketch map of the geology of Mishina Gora (after Masaitis).

Access: Mishina Gora is located in a relatively-well-populated area just 25 km southeast of the city of Gdov, near the Estonian border at Lake Peipus (Chudskoye Ozero).

Reference
Masaitis, V. L., 1976, Astroblemes in the USSR. *Int. Geol. Rev., 18,* 1249–1258.

Morasko Craters

Poland

Lat/Long: N52° 29′, E16° 54′
Diameter: 0.1 km
Age: 0.01 Ma
Condition: Fresh

The Morasko craters were first recognized in 1957, when a connection was made between the earlier discovery of several meteorites in the region and the presence of seven (now eight) depressions, which had been assumed to be glacially-formed lakes. The first of the Morasko meteorites, which are coarse octahedrites, was found in 1914 during the First World War, when it was encountered during the digging of a rifle pit. A total of nearly 20 meteorite fragments have now been found, mostly from farmlands in the area. The largest weighs 78 kg. While the presence of a shower of meteorites of the same type in their surroundings is evidence that the craters might be meteoritic in nature, the proof came from a study of microscopic particles in the ground, which have been found to include large numbers of meteoritic particles, especially near the largest crater and to the north, where the larger meteorite finds apparently were concentrated. The meteoritic particles include both spherules and irregular fragments.

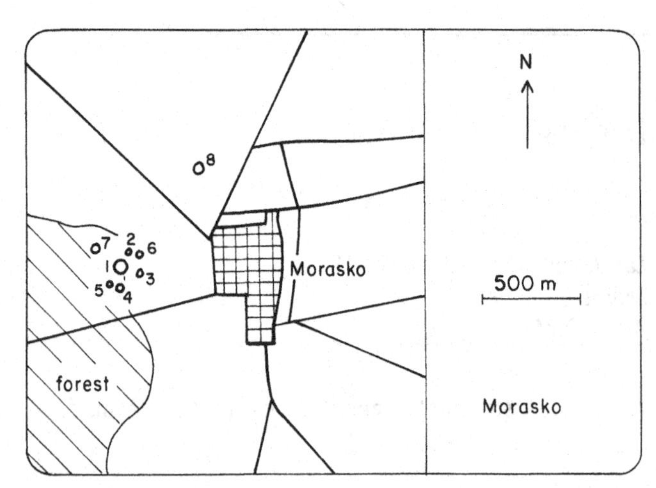

MORASKO: Location map of the craters at Morasko (after Kuzminski).

Seven of the craters are located in a beech forest and six of them contain shallow lakes. The eighth is located about 1 km to the northeast of the main group, and is now difficult to recognize, having been in a farmer's field and lost to continual plowing. It has been suggested that there may have been others that were similarly destroyed by farming activity. Apparently (though detailed descrip-

MORASKO: The largest crater at Morasko, Crater 1 (courtesy of H. Kuzminski).

MORASKO: Crater 3 at Morasko, looking north (courtesy of H. Kuzminski).

MORASKO: Crater 4 at Morasko, looking southward (courtesy of H. Kuzminski).

tions are missing) most or all of the meteorites were found in the area to the north and northeast of the main crater group, suggesting that the configuration was probably similar to that of the Sikhote Alin craters (Chapter 8), with the craters in an 'ellipse of fall' that has the larger craters at one end and the smaller meteorite fragments (those that were decelerated sufficiently to land short of the craters) at the other.

The different craters are described, using designations given by Korpikiewicz in the reference given below, as follows:

Crater 1
The largest crater has a diameter of 100 m and a mean depth of 13 m. It is lake-filled and surrounded by trees. The depth of the water varies, averaging about 2 m. The inner crater wall has a slope of about 18°. The northern rim is about 10 m higher than the southern rim, a pattern that has been reported to be typical of all of the craters.

Crater 2
This crater is 25 m in diameter and 3 m deep. It is not permanently water-filled.

Crater 3
The second largest crater, Crater 3 is 63 m in diameter and 5 m deep. It is lake-filled.

Crater 4
Also lake-filled, Crater 4 has a diameter of 35 m and a depth of 4.5 m.

Crater 5
This is the smallest crater, with a diameter of 15 m and a depth of only 0.9 m. It is usually not water-filled.

Crater 6
Similar in diameter to Crater 2, this one is somewhat shallower; its depth is 2.5 m and its diameter is 24 m.

Crater 7
About 300 m to the northwest of the main group, Crater 7 is 50 m across and 4.9 m deep.

Crater 8
While all of the above craters are located in woods, this crater is out in the open fields, about 1 km to the northeast. Its diameter is 35 m and its depth is 4.5 m.

Access: The Morasko craters are located just west of the village of Morasko, which is 9 km north of the city of Poznan, in the valley of the Warta River. They are preserved as the Morasko Meteorite Sanctuary.

References

Classen, J., 1978, The meteorite craters of Morasko in Poland. *Meteoritics, 13,* 245–255.
Korpikiewicz, H., 1978, Meteoritic shower Morasko. *Meteoritics, 13,* 311–326.
Kuzminski, H., 1980, The actual state of research into the Morasko Meteorite and the region of its fall. *Bull. Astron. Inst. Czech., 31,* 228–230.

Obolon

Ukraine

Lat/Long: N49° 30′, E32° 55′
Diameter: 15 km
Age: 215 Ma
Condition: Buried

The Obolon structure is a circular basin near the northeast edge of the Ukrainian shield. It lies buried under about 270 m of Cenozoic and Cretaceous sedimentary layers. Under these post-impact deposits are impact deposits, including breccias in which shock-deformation effects are found, including planar features in quartz grains and shatter cones. A hole was drilled at the center of the feature; a depth of 900 m was penetrated which reached deep-seated breccias and impact glass.

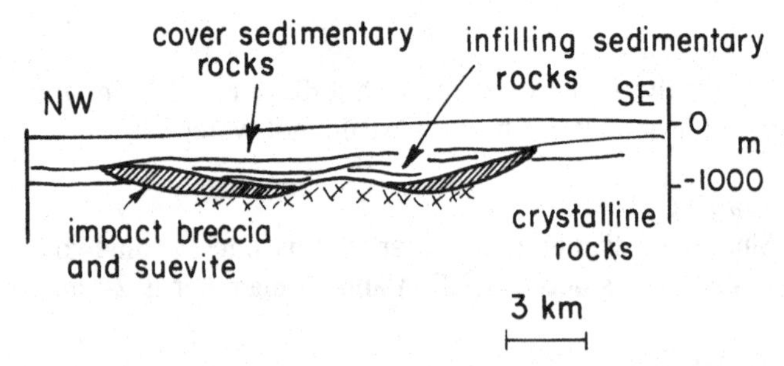

OBOLON: Profile of the Obolon structure (after Masaitis *et al.*).

Access: The Obolon structure is near the town of Obolon in the Poltava District, about 200 km southeast of Kiev, near the Dnieper River.

References

Masaitis, V. L., Danilin, A. N., Karpov, G. M. and Raykhlin, A. I., 1976, Karla, Obolon, and Rotmistrovka astroblemes in the European part of the USSR. *Doklady Earth Sci., 230,* 48–51.
Valter, A. A., Gurov, Y. P., and Ryabenko, V. A., 1978, The Obolon fossil meteorite crater (astrobleme) on the northeast flank of the Ukrainian Shield. *Doklady Earth Sci., 232,* 37–40.

Puchezh–Katunki

Russia

Lat/Long: N57° 06′, E43° 35′
Diameter: 80 km
Age: 220 Ma
Condition: Eroded, mostly buried

For nearly 100 years a large oval-shaped geological puzzle existed in central Russia, near the villages of Puchezh and Katunki. It lay mostly below ground; above it the Volga River flowed, Russian farmers farmed and geologists scratched their heads. Over the years a variety of different hypotheses were put forward to explain the strange geology of the structure, ranging from salt tectonics to ancient proluvium. Interest was high because of the possibility of mineral and oil deposits. Many holes were drilled, extending down to more than 1000 m, and gradually a clear picture of the Puchezh–Katunki Disturbance emerged. By 1964 it was suggested that the structure might be of impact origin and subsequent studies have supported that idea.

The target rock was a Triassic sedimentary series, about 2 km thick, that overlay Archean crystalline rocks. Drill cores show that the impact produced a central uplift, consisting of disturbed crystalline rocks, and a surrounding

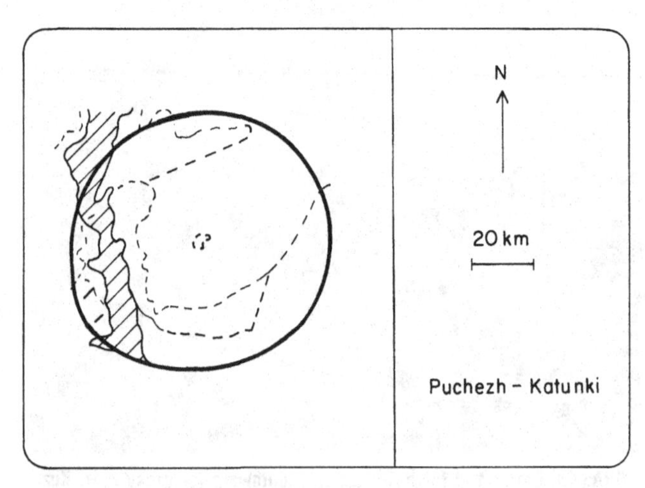

PUCHEZH–KATUNKI: Sketch map of the Puchezh–Katunki structure.

trough, partly filled with impact breccia and suevite, all covered with post-impact clays and sands. The summit of the central peak is about 2 km higher than the surrounding ring-shaped depression and it is about 20 km in diameter at the base. The top has a central depression about 4 km in diameter and 600 m deep. The upper rocks of the central uplift show various degrees of shock metamorphism.

Access: The Puchezh–Katunki structure is about 400 km east of Moscow, just north of the city of Gorkiy. It is centered on the village of Tonkovo.

References

Firsov, L. and Kieffer, S., 1973, Concerning the meteoritic origin of the Puchezh–Katunki Crater. *Meteoritics, 8*, 223–244.

Masaitis, V. L. and Mashchak, M. S., 1990, Puchezh–Katunki astrobleme: structure of central uplift and transformation of composing rocks. *Meteoritics, 25*, 383.

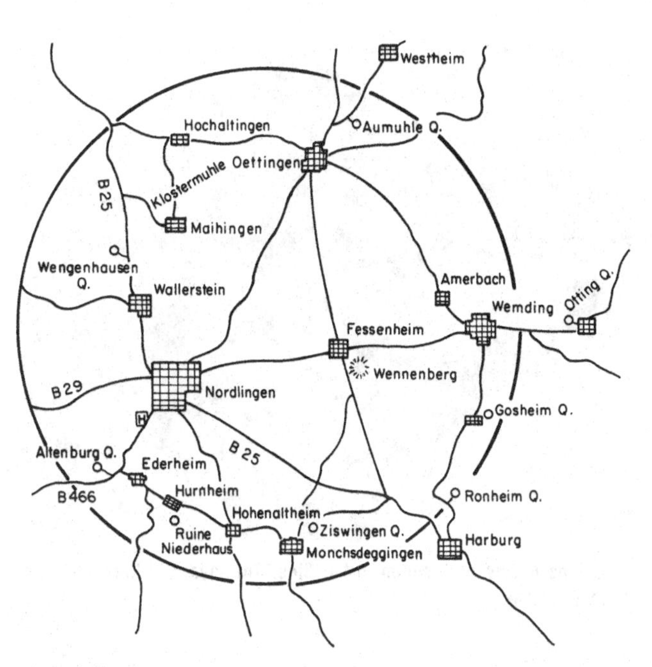

RIES: Map of the Ries basin, showing roads, towns and quarries mentioned in the text.

RIES: The Ries basin is fertile farm country, with a flat floor that is occasionally interrupted by isolated hills that represent parts of the central uplift. This view shows the Wennenberg Hill from Fessenheim.

Ries

Germany

Lat/Long: N48° 53′, E10° 37′
Diameter: 24 km
Age: 15 Ma
Condition: Eroded, largely covered

No-one with an interest in meteorite craters should visit Europe without a trip to the Ries basin. It is a spectacular example of a huge crater of considerable age, one that shows excellent outcrops of megabreccia and impact melt, and one that produced tektites that were jetted 500 km away. Besides, the residents of the area of the Ries have built a magnificent museum devoted to the crater. It is one of the best special-purpose science museums in the world and stands as a monument to the scientific integrity and imagination of the residents who built it and to the scientists who solved the long-standing mystery of the Ries.

Geographically, the Ries is a large, flat, circular valley surrounded by a rim of low hills that merge outward into the surrounding hilly Swabian Alb. The flat floor of the basin is an anomaly – hence its characterization as a 'kessel' ('kettle' in English). The old walled city of Nordlingen lies in the basin. Its main church is built primarily of what is now recognized to be impact breccia; the stone is beautiful and distinctive. Other small towns dot the valley and the majority of its floor is made up of prosperous farms.

There is a large literature on the Ries structure. A complete bibliography would list over 100 references. This description is a bare outline of the characteristics of this important feature. The two booklets listed at the end of this essay are more complete guides to its geology. The

RIES: The quarry at Gosheim in the Ries. The layers of limestone are almost vertical.

RIES: The quarry at Aumuhle in the Ries, now inactive, showing the weathered suevite wall.

RIES: The wall of the quarry at Otting in the Ries, showing a good view of the thick layer of suevite.

RIES: The walls of St. George's church in Nordlingen in the Ries were fashioned from suevite.

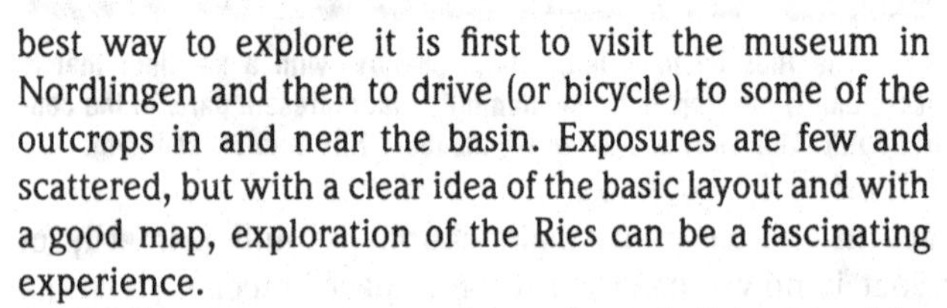

best way to explore it is first to visit the museum in Nordlingen and then to drive (or bicycle) to some of the outcrops in and near the basin. Exposures are few and scattered, but with a clear idea of the basic layout and with a good map, exploration of the Ries can be a fascinating experience.

The Ries impact occurred onto Jurassic limestone (the Malm formation), and penetrated into a basement of crystalline rocks. It produced a wide variety of impact-induced phenomena: tektites, shattered rocks, rock striations, various kinds of breccias, shatter cones and other shock-induced structures, such as planar features in quartz grains, ejected blocks (called 'Reuter blocks' and found many kilometers from the crater), suevite and glass bombs (called 'Flädle' and usually contained within the suevite). These different products of the impact are generally located in a radial pattern, with the Reuter blocks outermost and the most-shocked rocks innermost. The outcrops in the basin are primarily of post-crater deposits, the crater having been filled with a lake for a considerable time. Quarries at and near the rim are the best places to see the impact rocks, while the central uplift can be discerned in the hills making up the U-shaped inner crystalline ring, which consists of uplifted and shattered granite, gneiss and other rocks from the crystalline basement. Of the breccias the colorful 'Bunte Breccia' is the most interesting, consisting of a catholic mixture of about every kind of possible and available rock, including pieces from all the sedimentary layers and the crystalline basement; green clays, brown coal, white limestones, and red fragments all contribute to its remarkable appearance.

The Ries suevite is, of course, also quite remarkable. This type of impact rock obtained its name here, where such excellent examples can be found. Suevite is a special kind of breccia with a glassy matrix and including fragments of rock, large and small glass bodies and montmorillonite. It usually occurs over the Bunte Breccia, so was deposited afterwards. The suevite in the outer parts of the structure and its environs is the most glassy and it often contains pan-shaped Flädlen. Drill cores show that under the central crater basin the suevite is less glassy, is thicker (up to 400 m thick) and has no Flädlen.

The following is a guide to a brief tour of the Ries structure. A more complete tour, taking a full two days, is described by Kavasch (see references). The exposures in many places are marked by special signs with a crater design and an arrow. Some are accompanied by explanatory placards and even picnic tables. The tour below starts at the historic walled city of Nordlingen.

Distance from Nordlingen (km)	Feature
0	Visit the St. George's church, which was constructed during the interval 1454–1490 entirely out of suevite. The town hall and parts of the city wall were also made of this glassy impact breccia.
7	Drive southwest through Kleinerdlingen on B466 to the Altenburg quarry, an old suevite quarry where most of the stone used in old Nordlingen's prominent buildings was obtained. Only a little suevite is left, mostly on the northern exposure wall.
	Return to B466 and cross to the southeast on a side road through Ederheim to Hürnheim. Just past the village turn south for about 1 km.
10	Ruine Niederhaus is the ruins of a twelfth-century fortress made from hewn suevite blocks.
	Return to the Hurnheim road and continue southeast through Hohenaltheim to Monchsdeggingen, where a left turn leads to Ziswingen.
20	Just east of Ziswingen is the Ziswingen quarry, an active and very large quarry of limestone. The limestone is shattered and pulverized in some places and lies in its original layer sequence in others. To visit this and other active quarries, permission must be obtained from the site managers.
	Continue east from Ziswingen to Harburg, cross the Eger River and turn north through Ronheim (an active quarry is here) to Gosheim.
37	Just east of the village is an abandoned quarry at the position of the crater rim. The Malm limestone beds are overturned. Partial folding and displacement of the beds also attest to the impact event. Fossils are abundant in the limestone here.
	Continue north to Wemding and turn east to Otting.
47	The Otting quarry provides the best outcrop of suevite in the Ries. The suevite layer is about 20 m thick and can be seen to overlie the Bunte Breccia here.
	Return to Wemding and turn northwest through Amerbach to Oettingen. Turn northeast towards Westheim and drive 3 km from Oettingen and turn right.
69	The Aumuhle quarry is inactive. It has a nice exposure of suevite overlying the Bunte Breccia.
	Return to Oettingen and go west past Hochaltingen, turn south to Maihingen and turn west again to Klostermuhle.
86	Walk across the bridge crossing the stream west of the Kloster and notice the outcrop of crystalline rock from the crater basement. It is mainly gneiss and shows extensive shattering. There is a nice small museum here, the Rieser Bauernmuseum, open mainly on weekends.
	Proceed north to Marktoffinger and turn south on B25; go 1.5 km and turn west onto a dirt road, which leads about 1 km to the Wengenhausen quarry.
91	The Wengenhausen quarry shows strongly-shattered basement rocks of granite, gneiss and amphibole with small

Distance from Nordlingen (km)	Feature
	dikes of fine-grained breccia. Above it lies fossil-rich limestones from the post-impact lake. Continue on B25 to Wallerstein.
95	A walk to the top of the Wallerstein cliff provides an excellent view of the entire basin. The hill is part of the inner crystalline ring of uplifted, disturbed basement rock.
100	Return to Nordlingen.

Access: The Ries basin is in southern Germany and easily reached by Autobahn from Munich or any other German city.

References

Chao, E. T., Huttner, R. and Schmidt-Kaler, H., 1978, *Principal Exposures of the Ries Meteorite Crater in Southern Germany*. Bayerisches Geologisches Landesamt, Munich, 84 pp.
Kavasch, J., 1985, *The Ries Meteorite Crater*. L. Auer, Donauworth, 87 pp.

Rochechouart

France

Lat/Long: N45° 50′, E0° 45′
Diameter: 23 km
Age: 186 Ma
Condition: Highly eroded

The Rochechouart structure lies in the peaceful and picturesque area of the Vienne Valley in central France. It is near the western edge of the Central Massif and the country is hilly, with a mixture of farms and woods. Just east of the center of the structure is a sixteenth-century castle-like manor house at the village that gives its name to the structure, Rochechouart. This graceful building is one of the most impressive structures in the world constructed (unknowingly) of impact breccias (another is the church in Nordlingen).

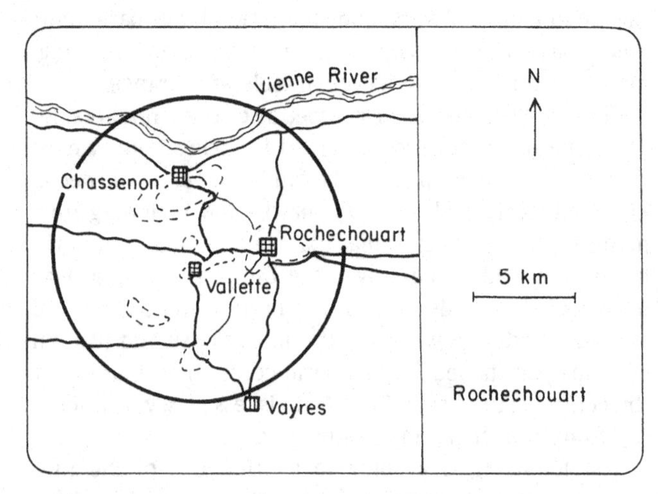

ROCHECHOUART: Map of the location of the Rochechouart structure.

ROCHECHOUART: The manor house at Rochechouart was built of impact breccia.

The impact structure does not present a distinctive topography. Its true nature, first proposed in 1967, was suggested on the basis of the presence of distinctive impact breccias, suevites and impact melt rocks, which are obvious in local quarries. The impact structure is located in old crystalline rocks, including granites, gneisses, and shists. It is best discerned by a few surface outcrops and by several quarries.

The crater is eroded almost entirely away. All that remains is the bottom of the crater, with its fall-back breccias and impact melts. The breccias show a gradient of degree of shock, with the most shocked and melted rocks near the center. Patches of breccia are found beyond the 12-km-sized central zone out to about 10 km from the center. Shatter cones are found in both shists and granites and quartz grains exhibit shock lamellae. Meteoritic contamination of the melt rocks indicates that the projectile was an iron meteorite.

ROCHECHOUART: The Meteoritical Society visits a quarry near Rochechouart, from which impact breccia was mined.

ROCHECHOUART: A piece of colorful impact breccia from Rochechouart.

Access: The Rochechouart structure is about 50 km west of the pleasant city of Limoges. The Vienne River crosses its northern boundary, the town of Rochechuart is just east of its center and the village of Vallette is about at its center.

References

Kraut, F. and French, B. M., 1971, The Rochechouart meteorite impact structure, France: preliminary geological results. *J. Geophys. Res., 76*, 5407–5412.

Lambert, P., 1977, Rochechouart impact crater: statistical geochemical investigations and meteoritic contamination, in *Impact and Explosion Cratering*, ed. D. Roddy, R. Pepin and R. Merrill, pp. 449–460. Pergamon Press, New York.

Rotmistrovka

Ukraine

Lat/Long: N49° 00′, E 32° 00′
Diameter: 2.7 km
Age: 140 Ma
Condition: Buried

The small basin called Rotmistrovka is a circular depression in granite. The target rock is intensely shattered in the center and there are examples of melt rock. Breccia overlies the basement rock to a depth of 60 m. Glassy bombs are found and the melt rock near the top of the crater floor is similar to suevite. Planar elements in quartz grains, as well as other microscopic evidence of shock effects, are found in fragments of rock encased in the glass. The basin is covered by sediments, including clay, sand and marl.

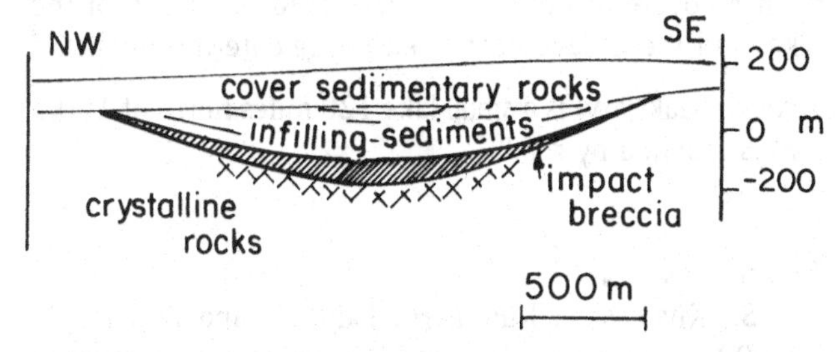

ROTMISTROVKA: Profile of the Rotmistrovka structure (after Masaitis *et al.*).

Access: The basin is near the town of Rotmistrovka, about 200 km southeast of Kiev.

References

Masaitis, V. L., Danilin, A. N., Karpov, G. M. and Raykhlin, A. I., 1976, Karla, Obolon, and Rotmistrovka astroblemes in the European part of the USSR. *Doklady Earth Sci., 230*, 48–51.

Saaksjarvi

Finland

Lat/Long: N61° 23′, E22° 25′
Diameter: 5 km
Age: 514 Ma
Condition: Eroded, lake-filled

Lake Saaksjarvi in southwestern Finland has been investigated geophysically, including gravity, magnetic, electromagnetic and radiometric surveys. All results indicate that the structure has the characteristics of an impact basin. A negative gravity anomaly, a lack of magnetic anomalies (which are common in the surroundings), and aeroelectromagnetic anomalies all agree. There are few outcrops, but drilling has revealed the presence of impact breccias. Drilling must be done in the winter, taking advantage of the ice cover of the lake. Suevite-like melt rocks have been recovered from glacial deposits to the south of the lake, and these have been used to age-date the impact.

Access: Saaksjarvi is a small lake 120 miles north of Turku and is reached by road.

References

Elo, S., Kivekas, L. Kujala, H., Lahti, S. and Pihlaja, P., 1990, Recent studies of Lake Saaksjarvi meteorite impact crater, Soutwestern Finland, in *Abstracts, Fennoscandian Impact Structures* ed. L. Pesonen and H. Niemisara, p. 44. Geological Survey of Finland.

Muller, N., Hartung, J., Jessberger, E. and Reimold, W., 1990, ^{40}Ar–^{39}Ar ages of Dellen, Janisjarvi and Saaksjarvi impact craters. *Meteoritics, 25*, 1–10.

Siljan

Sweden

Lat/Long: N61° 02′, E14° 52′
Diameter: 55 km
Age: 368 Ma
Condition: Eroded, partly lake-filled

The very large Siljan structure is probably the best-known impact basin in Sweden, but not because of its astronomical origin. Its fame, probably temporary, has another

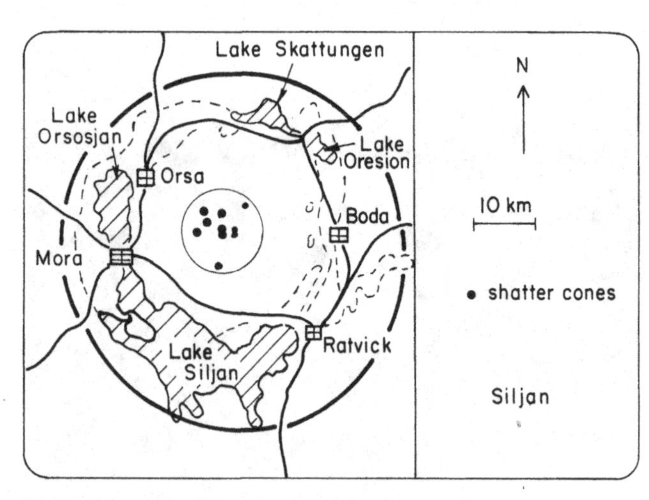

SILJAN: Map of the Siljan structure (after Svensson).

explanation, as described below. The geology of the structure is fairly typical for its size. The seismic data and drill cores show the presence of a central uplift and a surrounding ring-shaped graben. The mild vertical expression of the topography shows the result of considerable erosion. There are shatter cones present, as well as other evidence of shock metamorphism, such as lamellae in quartz grains. Impactite, internal megabreccia and fallout breccia have been identified and mapped, the last extending to several kilometres beyond the original rim.

The fame of the structure derives from the fact that it has been the source of a remarkable geological and economical gamble. In the late 1970s the astrophysicist Thomas Gold promoted the idea that there might be a vast reservoir of natural gas trapped below the Swedish granite, seeping up from the mantle, where Gold believed it had been accumulated during the formation of the Earth. He predicted that a drill core taken down through the Siljan structure might provide the easiest access to this Gold mine of abiogenic hydrocarbons, as the impact might have punctured the crustal rock deeply enough that mantle gasses could seep up and be detected and tapped. The Swedish State Power Board took the idea seriously and began a deep drilling project in 1987. Both public and private investors became involved. In one prospectus, issued by Anathema Oil, one of the holding companies, a total of 800 billion cubic meters of natural gas was advertised as possibly lying below Siljan, with a commercial value of about $40 billion (US). By 1989 drilling had reached 6800 meters, with no sign of the promised gases, and the project was, at least temporarily, stopped.

Access: Siljan is in central Sweden, about 250 km northwest of Stockholm. It is easily accessible, being partially filled by a scenic resort lake. The town of Mora is at the western rim.

References

Bottomley, R. J., York, D. and Grieve, R. A. F., 1978, ^{40}Ar–^{39}Ar ages of Scandinavian impact structures: L. Mien and Siljan. *Contrib. Minerol. Petrol.*, *68*, 79–84.

Kerr, R. A., 1990, When a radical experiment goes bust. *Science*, *247*, 1177–1179.

Svensson, N. B., 1971, Probable meteorite impact crater in Central Sweden. *Nat. Phys. Sci.*, *229*, 90–92.

Soderfjarden

Finland

Lat/Long: N63° 02′, E21° 35′
Diameter: 6 km
Age: 550 Ma
Condition: Eroded, covered

The Soderfjarden structure lies on the western coast of Finland. In recent centuries it has gradually emerged after having been beneath the waters of the Sea of Bothnia. In the fourth century it was entirely sea-covered, but by the thirteenth century it had risen to be a circular bay. It was recognized as a possible circular feature of interest in the 1970s from satellite images. Subsequent investigation of the rocks by drilling and of the gravity field confirmed its impact origin. The gravity models show a circular depression surrounding a central uplift, which is 40 m below the present surface. The rocks in the cores from the central uplift consist of crushed rock that is chemically identical to the country bedrock, which is Vassa granite. Quartz grains in the crushed rock show characteristic planar elements with two orientations.

Access: The Soderfjarden structure is south of the city of Vassa on the western coast of Finland and can be reached by road. None of the crater's rocks are exposed at the surface.

Reference

Lehtovaara, J. J., 1991, Soderfjarden, Western Finland, a Cambrian-filled meteorite crater 100 km west of Lappajarvi, in *Abstracts, Fennoscandian Impact Structures*, vol. 29, ed. L. Pesonen and H. Niemisara, p. 29. Geological Survey of Finland.

Steinheim

Germany

Lat/Long: N48° 02′, E10° 04′
Diameter: 3.4 km
Age: 15 Ma
Condition: Eroded, partially exposed

The Steinheim Basin is a unique feature in its setting. It lies in the scenic Swabian Alb of southern Germany, surrounded by flat, woodsy uplands, dotted with picturesque towns. The Steinheim Basin interrupts this low-relief hilly area as an abrupt, circular, flat-floored basin, lying about 90 m below its surroundings. The center of the basin is a central peak, rising about 50 m above the basin floor, with a diameter of about 900 m. The village of Steinheim extends north from the central peak (which is a park) and the village of Sontheim lies at the southern edge of the basin.

The surrounding rock strata are limestones and marls. The floor of the basin, on the other hand, is made up of lake deposits from the Tertiary and Quaternary. The exposures on the central peak show limestones, marls, clays, and sandstones of formations that are found in the surroundings, but at greater depths, indicating that the central peak represents an uplift of some 200 m or so.

Steinheim was the locale of the first discovery of shatter cones, although at the time (1905), their significance was not understood. The structure was thought to be a cryptovolcanic feature by most geologists until the time of extensive exploration by drilling in the 1960s and 1970s, which demonstrated its impact origin.

STEINHEIM: A photograph of the map that stands near the southern entrance to the Steinheim basin, inviting pedestrians to take advantage of walks through the structure.

STEINHEIM: The Steinheim basin from the Burgstall, with the village of Sontheim in the foreground and the central uplift in the center. The town of Steinheim is in the upper left. A sign with a geological profile of the basin is mounted at the lower right.

STEINHEIM: The Steinheim basin looking south from the hill of the central uplift. The Burgstall rises above the village of Sontheim.

In addition to shatter cones found in both the central uplift and the outer walls, there is fallback breccia exposed, especially at the south edge of the structure and in the northeast. The breccia shows quartz grains with planar features, indicating shock metamorphism. Strata in the central uplift show a high degree of fracturing and are tilted by 30–60°, while cores demonstrate that they can be classified as a megabreccia. Shock effects in them decrease with depth.

The 'Burgstall' is a linear hill that makes up a wall corresponding to the southern edge of the basin. It is a block of upthrusted rock in which breccia and shatter cones are evident. The 'wall' is about 200 m wide and extends part-way around the basin.

Steinheim is only 40 km west-southwest of the Ries Basin and this proximity plus its similarity in age have led to the suggestion that the two were formed simultaneously, either by a double asteroid or by one that split into two pieces upon encountering the Earth's atmosphere.

The following provides a pleasant itinerary for a walk through the Steinheim Basin. Begin the tour at the 'Burgstall', where there are large signs and a description of the structure.

Distance from the Burgstall (km)	Feature
0.0	Behind the sign is an exposure that shows both breccia and some shatter cones. This hill is the breached 'wall' that extends around this portion of the basin. It is made up of steeply tilted and deformed strata and breccia. A path leads to the top, from which excellent views of the basin are obtained. Proceed north along the road to Steinheim, passing through the village of Sontheim. The flat part of the basin is largely farmland, underlain by Quaternary deposits and Tertiary lake beds.
1.0	Passing the central uplift to the right, notice its rounded, conical shape. At the stairs and path to the right, proceed up the central hill to the summit, where there is a memorial monument.
1.5	From the top it is possible to see the outer walls of the basin in several directions and, just in front of it, the hills that form the inner 'wall'. Continue east on a path to where it divides, where a path to the south is taken.
1.8	The view opens up as the path emerges from the woods of the park. Outcrops include fractured examples of limestones and other rocks from formations normally found hundreds of meters deeper.
2.0	Descend into farmland due north of Sontheim and follow paths to the southwest and south back to the starting point.

STEINHEIM: From the top of the Burgstall looking west along the rim of the basin.

Access: The Steinheim Basin is easily accessed from the Autobahn. It is 7 km west of Heidenheim.

Reference

Reiff, W., 1977, The Steinheim basin: an impact structure, in *Impact and Explosion Cratering*, ed. D. Roddy, R. Pepin and R. Merrill, pp. 309–320. Pergamon Press, NY.

Ternovka

Russia

Lat/Long: N48° 01′, E33° 05′
Diameter: 12 km
Age: 330 Ma
Condition: Eroded

The Ternovka structure was originally thought to be a breccia pipe, but the discovery of shock metamorphism, including the minerals coesite and stishovite, showed in 1980 that it is undoubtedly an impact structure. The crater cuts into shists and other metamorphic rocks of the Ukrainian shield. Breccias are found in outcrops and in cores. There is evidence of a central uplift and both tagamite and suevite are located near the center.

Access: The Ternovka structure is located just north of the city of Krivoy Rog.

Reference

Masaitis, V. L., Mashchak, M. S. and Sokolova, I. Y., 1980, High-pressure silica phases in the Ternovka astrobleme. *Doklady Earth Sci.*, *255*, 164–167.

Vepriaj

Lithuania

Lat/Long: N55° 06′, E24° 36′
Diameter: 8 km
Age: 160 Ma

The Vepriaj structure is in east-central Lithuania, north of Vilnius.

Reference

Grieve, R. A. F., 1991, Terrestrial impact: the record in the rocks. *Meteoritics*, *26*, 175–194.

Zapadnaya

Ukraine

Lat/Long: N49° 44′, E29° 00′
Diameter: 4 km
Age: 115 Ma

Zapadnaya is in central Ukraine, southwest of Kiev.

Reference

Grieve, R. A. F., 1991, Terrestrial impact: the record in the rocks. *Meteoritics*, *26*, 175–194.

Zeleny Gai

Ukraine

Lat/Long: N48° 42′, E32° 54′
Diameter: 2.5 km
Age: > 140 Ma
Condition: Buried

The Zeleny Gai structure was discovered as a circular gravity low, located near the town of that name. Drilling revealed that the gravity low resulted from the presence of a crater-shaped basin in the underlying crystalline rocks. The overlying rocks are Cenozoic sediments, but the basin's rocks were found to include breccias (of a wide variety of characteristics) and melt rocks. Evidence of shock metamorphism was found in the mineral grains of the breccias.

Zeleny Gai is only about 40 km from the Boltysh and Rotmistrovka impact structures and it was once suggested that they may have originated from the same incoming body. The younger age of the Boltysh structure, however, seems to eliminate it from this possible multiple-impact hypothesis.

Access: The structure is located near the town of Zeleny Gai, about 250 km south-southeast of Kiev.

Reference

Valter, A. A., Bryanskiy, V. P., Ryabenko, V. A. and Lazarenko, Y. Y., 1976, The Zeleny Gai impact structure. *Doklady Earth Sci.*, *229*, 34.

7

Impact structures of Africa

Amguid

Algeria

Lat/Long: N26°05, E4° 23′
Diameter: 0.45 km
Age: <0.1 Ma
Condition: Fresh, exposed

The relatively young Amguid crater is located in Lower Devonian sandstones. It is a nearly perfectly-circular basin with a raised rim that reaches about 30 m above the crater floor. The center of the floor is covered with fine, white eolian deposits and the rim and outer slopes are covered with loose ejecta, made up of sandstone fragments. The

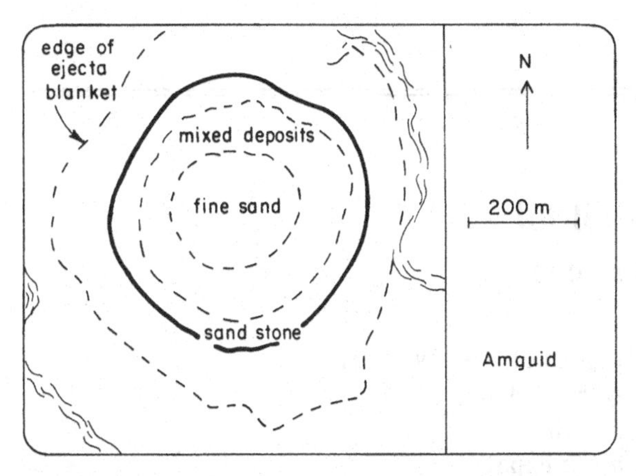

AMGUID: Geological sketch map of the Amguid crater (after Lambert *et al.*).

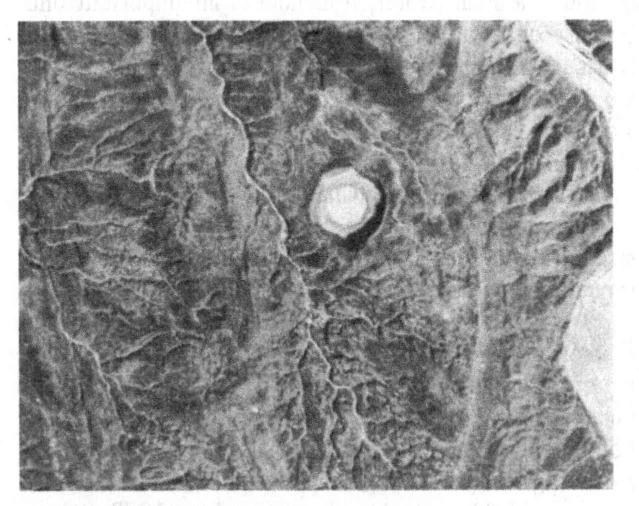

AMGUID: Aerial photograph of the Amguid crater. At the right is the Oued Tafrakrek Canyon (courtesy of J. F. McHone, Jr.).

ments. The otherwise nearly-horizontal beds are upturned near the rim, becoming nearly vertical at the top of the rim. On the north-northwest and south-southeast parts of the rim the beds are overturned. Planar elements are observed in some quartz grains. Geologically, Amguid is very similar to the somewhat larger Barringer (Chapter 2) and Wolfe Creek (Chapter 5) craters.

Access: The Amguid crater is not located near a major highway. It is about 1 km west of the Oued Tafrakrek Canyon and about 90 km west-southwest of an old fort named Amguid, the nearest named feature on the map. The Oasis of In Salah, where there is an airport, is 225 km to the northwest.

Reference

Lambert, P., McHone, J. F., Dietz, R. S. and Houfani, M. 1980, Impact and impact-like structures in Algeria. I. Four bowl-shaped depressions. *Meteoritics, 15*, 157–179.

Aouelloul

Mauritania

Lat/Long: N20° 15′, W12° 41′
Diameter: 0.4 km
Age: 3.1 Ma
Condition: Exposed

Although a small crater, Aouelloul is an important one because of its young age and its tektite-like impact glass. It occurs in modestly-dipping beds of quartzite and sandstone and it shows many of the characteristics typical of small craters. Its most unusual feature is the presence of abundant solid fragments of impact glass, different from those found at Henbury (Chapter 5) or Wabar (Chapter 8), for example, which are frothy. The glass shows flow structure and is chemically almost identical to the target sandstone. Minute enhancements of iron, nickel and cobalt suggest that the glass is contaminated with target meteorite material. The glass is found in and around the crater, especially just outside the rim to the southeast. Its peculiarity and the relative shallowness of the crater have both been attributed to such anomalous circumstances as a low-angle impact or a glass projectile.

A remarkable coincidence occurred in 1973, when scientists discovered a small stony meteorite adjacent to the crater, near the area of highest glass fragment density. Subsequent analysis, however, showed that the juxtaposition must be coincidental. The meteorite had a terrestrial lifetime of only 300 000 years, $^{1}/_{10}$th the age of the crater.

A more remote meteorite, but one that was said to be located in the same area of the desert, is the probably mythical Chinguetti Meteorite. Early explorers into the Adrar Desert brought back a chunk of what turned out to be a genuine iron meteorite, along with the claim that it was broken off from a huge mass, as large as a house and weighing an estimated 10^5 tons. Although subsequent searches were made, the Chinguetti Meteorite was never found. The discoverers suggested that it had been buried by the shifting sands, and that it lies there still, waiting to be uncovered and rediscovered. In spite of the unliklihood that such a large meteorite could survive terrestrial impact in one piece, romantics still hold out hope that the Chiguetti iron, largest meteorite in the world, will turn up again.

It has been suggested that the Tenoumer structure, also in Mauritania, may be contemporaneous with Aouelloul, and that they may have been part of the same fall.

Access: The crater of Aouelloul is only 50 km southeast of the city of Atar, where there is an airport.

References

Chao, E. C. T., Merrill, C. W., Cuttitta, F. and Annell, C., 1966, The Aouelloul crater and the Aouelloul glass of Mauritania, Africa. *Trans. Am. Geophys. Union, 47*, 144.

Fudali, R. and Cressy, P. J., 1978, Investigation of a new stony meteorite from Mauritania with some additional data on its find site: Aouelloul crater. *Earth Planet. Sci. Lett., 30*, 262–268.

Bosumtwi

Ghana

Lat/Long: N6° 32′, W1° 25′
Diameter: 10.5 km
Age: 1.3 Ma
Condition: Water-filled

The Lake Bosumtwi crater is a spectacular feature of the near-coastal African landscape. It is an unusual impact crater in several ways: it is lake-filled, it lies hidden in a

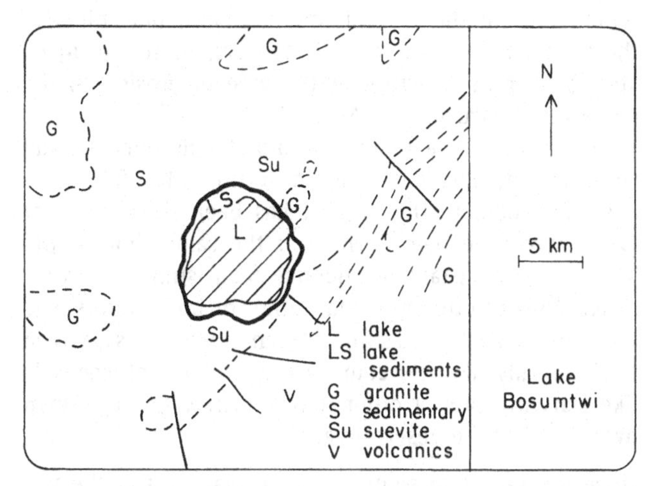

BOSUMTWI: Geological sketch map of the Bosumtwi crater (after Jones *et al.*).

jungle-like forest, its rim is well preserved and well-defined, and it is responsible for one of the Earth's few tektite strewn fields. The Ivory Coast tektites have been identified without question as having been produced by the Bosumtwi event, in much the same way that the Moldavites were produced by the Ries. During the fraction of a second before the meteorite hit the ground, the shock-compressed air layer at its front seared the earth, liquefying a layer of it and jetting it out across the landscape to a distance of some 300 km to the west. Now, a million years later, button-sized tektites are still found on the land and millions of microscopic-sized tektites can be recovered from the nearby ocean sediments.

Lake Bosumtwi is a nearly circular lake, surrounded by a rim that rises 250 to 300 m above lake level. The rim is broad and is skirted by a shallow depression and an outer rim 20 km in diameter. The lake has a maximum depth of 80 m. There is a terrace on the steep crater walls about 125 m above the present lake level, probably formed when the lake level reached this height at some point in its past. This would be the maximum water level height, because if it went higher it would spill out through the lowest point on the rim.

Impact breccias are found in many places at Lake Bosumtwi. One type of breccia is made up of the adjacent kind of country rock and can be found on the crater rim. Another type of breccia, consisting of coarse rocks in a fragmentary matrix of the same composition, is found down along the crater walls and on the outer ridge. These rocks were formed by being shattered and then cemented together more or less in place. A third kind of breccia consists of a mixture of many rock types and can be found primarily on the outer ridge. It is estimated that there is a layer of this highly-mixed breccia at least 20 m thick.

A rarer and more special kind of rock is the Bosumtwi suevite, a glassy breccia rather similar to the Ries suevite. It is found in a few stream valleys, especially at the south, where the Ata stream cuts through the rim towards the lake. The suevite is grayish-purple in color, with small rock fragments of a variety of color and origin, together with yellowish-brown bubbly glass. The suevites have been used to establish the age of the crater, using both Potassium – Argon and fission track techniques.

Shock – metamorphism effects are well documented at Lake Bosumtwi. Quartz grains show shock lamellae. Coesite is present, as are nickel – iron spherules in the suevite. In fact, the chemical evidence of these spherules, plus the anomalous nickel – iron content of the tektites, has led to the conclusion that the projectile was an iron meteorite, estimated to have been about 300 m in diameter and to have weighed about a hundred million tons.

Access: Lake Bosumtwi is 30 km southeast of the city of Kumasi, which is served by air from Accra, a major international port. From Kumasi to the lake, one follows the road to Bekwai and turns to the left just past Bekwai towards Morontuo. A secondary road off of this road leads to the south side of the lake. The north and east sides also have road access. It is advised that a local guide be engaged to find the best way through the usually-wet and rather deep forest. There are several small villages near the lake.

References

Jones, W. B., Bacon, M. and Hastings, D. A., 1981, The Lake Bosumtwi impact crater, Ghana. *Geol. Soc. Am. Bull., 92,* 342–349.

Saul, J. M., 1964, Field investigations at Lake Bosumtwi (Ghana) and in the Ivory Coast tektite strewn field. *Nat Geogr. Soc. Res. Rep., 1964,* 201–212.

BP Structure

Libya

Lat/Long: N25° 19′, E24° 20′
Diameter: 2.8 km
Age: <120 Ma
Condition: Exposed, deeply eroded

The BP structure was named after an oil company. The first report of its impact origin was made by an employee of the company. It lies in the Cyrenaican desert of eastern

Libya. The rocks of the area are all Nubian sandstone and much of the surroundings of the structure are sand-covered. BP stands out in the form of two concentric rings of low hills that surround a central rise that is somewhat asymmetrically eroded. The outer ring, 2.8 km in diameter, rises between 10 and 20 m above the plain, while the inner ring, 2 km in diameter, has a height of about 30 m. The central peak or uplift has a maximum height of about 40 m. The southeast half of the central peak is largely eroded away.

The sandstone beds in the outer ring show an inward dip averaging about 10°, while the inner ring's beds dip outward more steeply, averaging about 30°. The beds exposed in the central peak are highly folded and distorted. Samples of sandstone show shock metamorphism in the form of lamellae in quartz grains.

The similar degree of erosion of BP with that of the Oasis structure, 80 km to the south, has suggested that the two objects may have been formed at the same time.

Access: The BP structure is in a remote part of the eastern Libyan desert. It is 165 km northeast of the Kufra Oasis, where there is an airport.

References

French, B., Underwood, J. and Fisk, E., 1974, Shock-metamorphic features in two meteorite impact structures, Southeastern Libya. *Bull. Geol. Soc. Am., 85*, 1425–1428.

Martin, A. J., 1969, Possible impact structure in southern Cyrenaica, Libya. *Nature, 223*, 940–941.

Oasis

Libya

Lat/Long: N24° 35′, E24° 24′
Diameter: 11.5 km
Age: <120 Ma
Condition: Exposed, deeply eroded

The Oasis structure, like the BP structure, was named after an oil company. It consists of a broken ring of steep hills, about 100 m high and forming a ring 5 km in diameter. Exterior to this ring is a much less pronounced ring-shaped feature about 11.5 km in diameter. The center of the structure is relatively flat. The morphology is that of a highly-eroded, large impact structure in which only the outer walls of the central uplift remain, plus hints of the inner walls of the rim. It appears, therefore, to be very similar in structure and degree of erosion to the Gosses Bluff structure in Australia.

The rocks in the inner ring dip steeply outward and show considerable folding and distortion. Most folds have axes tangential to the ring. It has been suggested that Oasis may have been formed at the same time as BP. There is circumstantial evidence (consisting of relative proximity and little else) that either or both craters may have produced the puzzling Libyan Desert Glass, which is chemically like the country rock but morphologically like Darwin Glass or other impact glasses and is strewn over much of the desert area.

Access: Oasis is in southeastern Libya, west of the volcanic Gilf Kebir Plateau and east of the Kufra Oasis.

References

French, B., Underwood, J. and Fisk, E., 1974, Shock-metamorphic features in two meteorite impact structures, Southeastern Libya. *Bull. Geol. Soc. Am., 85*, 1425–1428.

Ouarkziz

Algeria

Lat/Long: N29° 00′, W7° 33′
Diameter: 3.5 km
Age: <70 Ma
Condition: Eroded

The Ouarkziz structure interrupts a large-scale arc of sedimentary rocks in the rocky desert of western Algeria. It consists of a circular ring of 100-m-high walls, broken by drainage to the south. A smaller half-ring lies just interior to the northern part of the main ring. There are uplifted and outward-dipping sedimentary beds in the main ring and brecciation in the inner ring, as well as folding. Quartz grains show planar features, indicating the presence of shock metamorphism in the rocks of the inner ring.

Access: Ouarkziz lies near the Algeria–Morocco border, adjacent to the Ouarkziz mountains. It is about 170 km north-northeast of Tindouf, where there is an airport.

Reference

Lambert, P., McHone, J. F., Dietz, R. S., Briedj, M. and Djender, M., 1981, Impact and impact-like structures in Algeria. II. Multi-ringed structures. *Meteoritics, 16,* 203–227.

Pretoria Salt Pan

South Africa

Lat/Long: S25° 24′, E28° 05′
Diameter: 1.1 km
Age: 0.2 Ma
Condition: Exposed, eroded

The Pretoria Salt Pan is a fairly well-preserved crater with a nicely-raised rim and containing a salty lake. The floor of the crater is 120 m below the rim, which rises 60 m above the surroundings. The country rock is granite, which is laced with small amounts of mafic and alkaline intrusive rocks. The rim shows uptilting of the granite and breccia consisting of granite fragments is found. There are both radial and concentric normal faults in and along the rim.

Although for years it was not known whether the Pretoria Salt Pan was an impact crater or whether it had a volcanic (or 'cryptoexplosive') origin, drill cores taken in 1989 showed a typical impact structure's vertical profile. Breccias were recovered that contained shock-metamorphosed quartz grains. A geophysical survey also confirmed the impact hypothesis when it detected a gravity anomaly that agreed with the sub-surface model of the crater based on the cores. There is a layer of post-impact lake sediments 90 m deep, underlain by 53 m of unconsolidated breccia made up of granitic sand and fractured granite boulders. Beneath this layer is fractured granite, which in turn is underlain by solid, undisturbed granite at a depth of about 200 m from the crater floor.

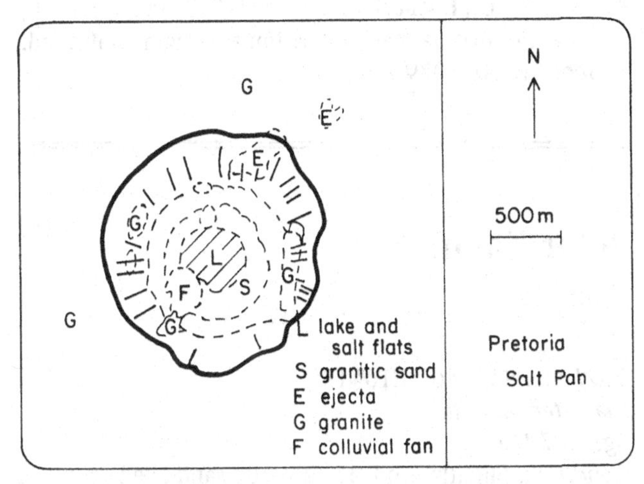

PRETORIA SALT PAN: Geological sketch map of the Pretoria Salt Pan (after Reimold).

PRETORIA SALT PAN: Aerial photograph of the Pretoria Salt Pan impact crater, taken from an elevation of approximately 500 m from a light aircraft. The photo was taken from the north, facing due south (courtesy of Dion Black).

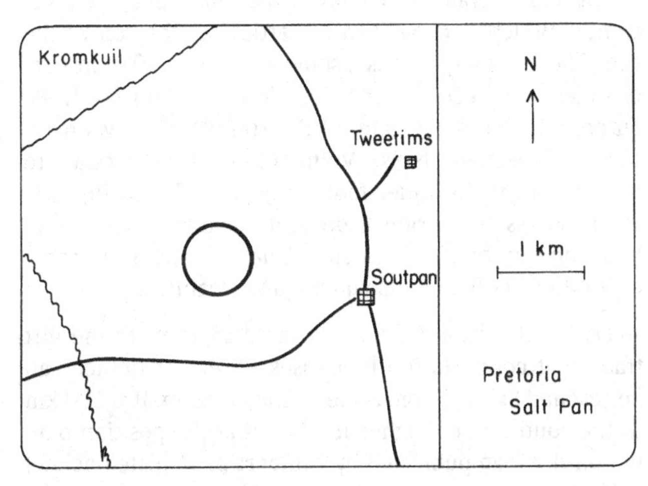

PRETORIA SALT PAN: Location map for the Pretoria Salt Pan.

Access: The Pretoria Salt Pan is about 40 km north-northwest of Pretoria, the capital of South Africa. It is in Bophuthatswana, just west of the township of Nuwe Eersterus, about 1 km from a main road.

References

Brandt, D. and Reimold, W. U., 1992, A structural and petrographic investigation of the Pretoria Saltpan impact structure. *Lunar Planet. Sci., 24,* 179–180.

Reimold, W. U., Koeberl, C., Partidge, T. C. and Kerr, S. J., 1992, Pretoria Saltpan crater: impact origin confirmed. *Geology, 20,* 1079–1082.

Roter Kamm

Namibia

Lat/Long: S27° 46′, E16° 18′
Diameter: 2.5 km
Age: 3.7 Ma
Condition: Slightly eroded and partly sand-filled

Roter Kamm has a conspicuous crater shape, especially clearly seen from space, from where its unusual morphology stands out against the fairly blank terrain of the southern Namib Desert. The country rocks consist of gneisses intruded by pegmatites and overlain by marble and sandstone. The crater has a rim that rises up to 150 m above the surroundings and a bowl-shaped basin that is mostly filled with sand and brecciated country rock. The raised rim is fractured with mostly radial faults. Impact melt rocks can be found on the rim. Ejecta in the form of fractured and deformed basement fragments have been found on the surrounding plain. High-pressure shock deformation in the form of planar elements and diaplectic glass are strong evidence for a meteoritic impact origin. Extensive hydrothermal activity after the impact is suggested by the unique melt breccias.

Access: Roter Kamm is in a barren and deserted area, not easily visited, though it is only 80 km north of the cities of Oranjemund and Alexander Bay at the mouth of the Oranje River. It lies within a Restricted Diamond Area.

ROTER KAMM: Aerial view of the Roter Kamm impact crater, taken from a helicopter at an elevation of approximately 150 m, looking towards the south (photo by R. McG. Miller, used through the courtesy of W. U. Reimold).

References

Fudali, R. F., 1973, Roter Kamm: evidence for an impact origin. *Meteoritics, 8,* 245–257.

Koeberl, C., Fredriksson, K., Gotzinger, M. and Reimold, W. U., 1989, anomalous quartz from the Roter Kamm impact crater, Namibia: evidence for post-impact hydrothermal activity. *Geochim. Cosmochim. Acta, 53,* 2113–2118.

Reimold, W. U., Miller, R. M., Grieve, R. A. F. and Koeberl, C., 1988, The Roter Kamm crater structure in SWA/Namibia. *Lunar Planet. Sci., 19,* 972–973.

Talemzane

Algeria

Lat/Long: N33° 19′, E4° 02′
Diameter: 1.75 km
Age: <3 Ma
Condition: Exposed and somewhat eroded

The Talemzane crater is a conspicuous circular crater with a raised rim and a deep (70 m) basin. It takes its scientifically-traditional name from a water reservoir 9 km to the southeast, but is locally known as Daiet el Maadna. It lies in nearly-horizontal beds of Eocene limestones. In shape it is similar to but more eroded than the Barringer crater (Chapter 2).

The crater wall shows outward-dipping beds of limestones, which become nearly vertical at the top of the rim. There are large blocks, some more than 10 m in size, of limestone ejecta on and outside of the rim crest. An outcrop on the outer flank of the south rim shows overturned limestone beds. Veins of impact breccia are common near the inner foot of the rim. The veins vary in thickness and orientation and include fragments of limestone up to 10 cm in size. Quartz grains from some of the breccia fragments show planar features.

Access: Talemzane Crater is located adjacent to the dirt track that connects the two oases, Laghouat (which can be reached by road from Algiers) and Guerara. It is 120 km to the southeast of Laghouat. Note that the position plotted on the map published by Lambert *et al.* (listed below) is incorrect.

Reference

Lambert, P., McHone, J. F., Dietz, R. S. and Houfani, M., 1980, Impact and impact-like structures in Algeria. I. Four bowl-shaped depressions. *Meteoritics*, *15*, 157–179.

Tenoumer

Mauritania

Lat/Long: N22° 55′, W10° 24′
Diameter: 1.9 km
Age: 2.5 Ma
Condition: Exposed, eroded

The Tenoumer crater has the form of a ring of rock encircling a plane that is at present at about the same level as the surrounding plain. The rim is 100 m high. Its geology has not been well-explored. The only unusual rocks so far reported are 'lavas' that occur just outside the rim. The basement rocks are granites and gneisses.

The Tenoumer structure was at first thought to be of volcanic origin. However, in 1970 contrary evidence was obtained in the form of the detection of shock lamellae in quartz grains from inclusions in the 'lava', which was shown to be impact melt, and by the recognition of lechatelierite and other evidence of shock effects. It is now accepted as an impact crater, though much is still be done in its study.

Access: The Tenoumer crater is in a remote part of the desert of central Mauritania, some 250 km from the nearest mapped road.

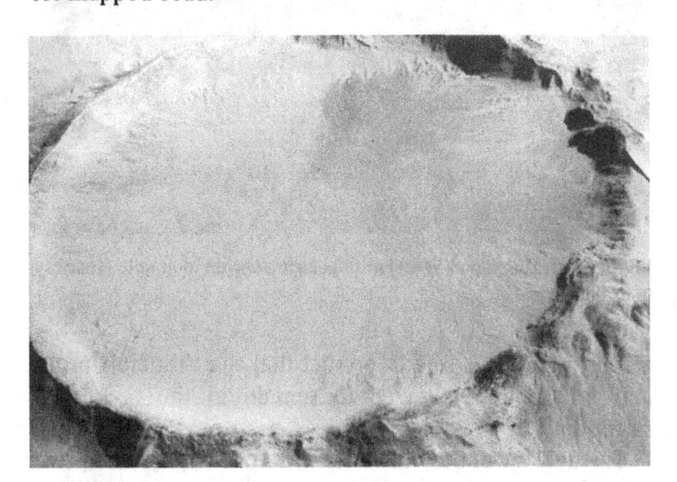

TENOUMER: Aerial photograph of the Tenoumer crater (courtesy of R. F. Fudali).

Reference

French, B. M., Hartung, J. B., Short, N. M. and Dietz, R. S., 1970, Tenoumer crater, Mauritania: age and petrologic evidence for origin by meteoritic impact. *J. Geophys. Res.*, *75*, 4396–4406.

Tin Bider

Algeria

Lat/Long: N27° 36′, E5° 07′
Diameter: 6 km
Age: <70 Ma
Condition: Deeply eroded

Tin Bider is a multi-ringed structure located in sedimentary rocks. It consists of at least three concentric, circular ridges, separated by troughs. The innermost area is an uplifted sandstone core, in which the exposed rocks are about 500 m higher than their normal position. This is surrounded by two limestone ridges with a diameter of 2 km, which is surrounded by another ridge 2.5 km from the center. A weakly-expressed outer ridge has a diameter of 6 km.

Tin Bider is highly unusual among impact structures of its size in one particular way. Instead of showing the usual fracturing and shattering, the target rocks are characterized by extensive soft folding, both at large and at small scales. Even limestone, a particularly brittle rock, shows ductile deformations of this sort. A typical cross-section through one of the ridges shows sedimentary beds that tilt 10° to 20° toward the center, with the lower beds undeformed and the upper beds softly folded and sometimes overturned or underturned. This peculiarity indicates that the deformation must have occurred in a confined state, such as would be the case far below the impact position, indicating that we are looking only at the roots of this structure. Also, it may have this unusual structure because of the fact that the target rocks included many clays and relatively soft rocks. The sandstones of the center are the only massive formations of the structure, and they have shattered and brecciated deformation of the more normal sort. Quartz grains in these sandstones show shock lamellae.

Access: Tin Bider is in central Algeria, about 265 km east-northeast of In Salah.

Reference

Lambert, P., McHone, J. F., Dietz, R. S., Briedj, M. and Djender, M., 1981, Impact and impact-like structures in Algeria. II. Multi-ringed structures. *Meteoritics, 16*, 203–227.

Vredefort

South Africa

Lat/Long: S27° 00′ E27° 30′
Diameter: 140 km
Age: 1970 Ma
Condition: Highly eroded, partly exposed

The Vredefort structure (sometimes called the Vredefort Ring or Dome) is one of the world's largest and oldest probable impact formations. It has taken a long time since its initial recognition as an unusual geological feature to reach a consensus regarding its origin. Even now there are geologists who regard the matter as open, if not settled in favor of a terrestrial origin. There has been a great deal of debate on the issue, much of it polite. For example, a comment published by a distinguished geologist who was responding to the paper that announced the discovery of radially-oriented shatter cones at Vredefort called the paper a 'bagatelle', called for bringing the discussion 'from the realms of astronomy back into the field of geology' and referred to the 'fantastic meteorite impact' theory. Furthermore, the published comments say that 'since the Vredefort structure is a factor in controlling the shape of the depository of the greatest known accumulation of gold

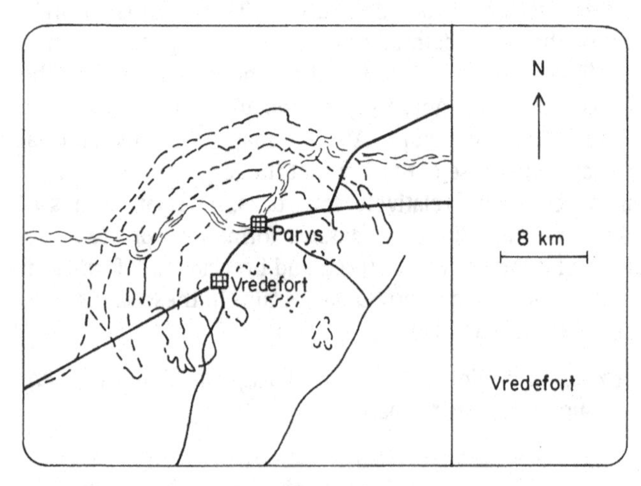

VREDEFORT: Location map of the Vredefort structure.

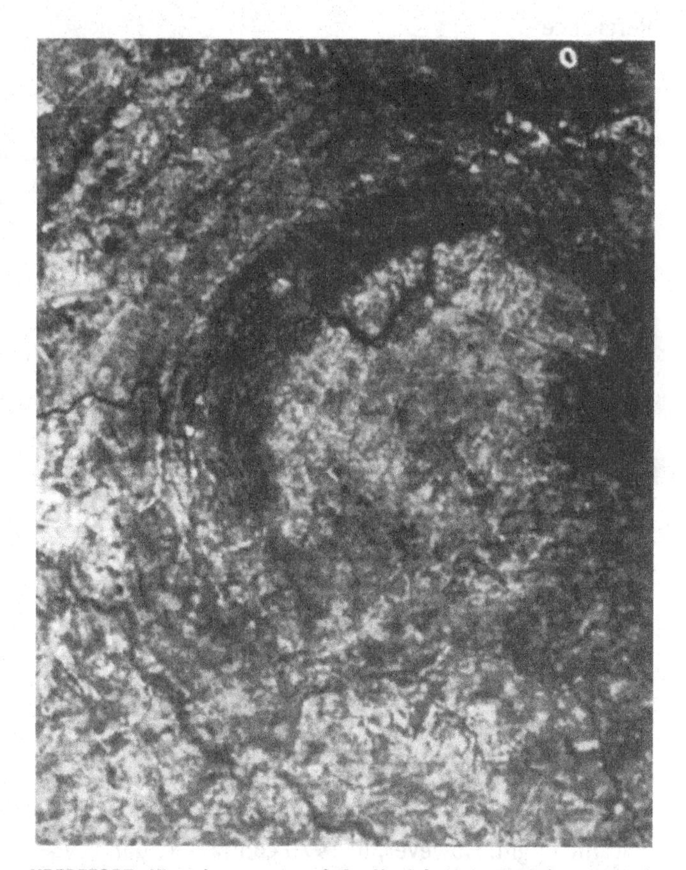

VREDEFORT: View from space of the Vredefort structure (courtesy of NASA).

VREDEFORT: Outcrop at Vredefort of pseudotachylite in granite (courtesy of E. Cheney).

in the world, it is safe to predict that the Vredefort problem will not be allowed to remain in the realms of astronomy'.

The geology of the Vredefort structure is quite complex. As at Sudbury (Chapter 3), much has happened to distort and erase the original formation. It is now a series of

VREDEFORT: Shatter-coning in quartzite at Vredefort (courtesy of E. Cheney).

concentric half-circles of outcrops at the boundary between two rock regimes, the Karoo System and the Transvaal System. Archean granite and shists, as well as several sedimentary systems, characterize the circular rings. The arguments in favor of an impact origin are the presence of a number of different shock-produced features, including shatter cones, coesite, stishovite, and pseudotachylite. As with the Sudbury complex, the Vredefort structure seems to occupy a special place in its geological setting (near the center of the huge Bushveld Complex), which is either a coincidence or an indication that it is related to the much larger-scale environs. If the latter possibility is true, then it can be used as an argument that Vredefort is endogenic. To explain it as some kind of explosive volcanic artifact, however, would necessitate positing that such a phenomenon could result in the extremely high pressures needed to form the observed shock-induced features.

Access: The Vredefort structure is easy to visit. It lies astraddle the main highway between Cape Town and Johannesburg, about 100 km southwest of the latter. The town of Vredefort and the city of Parys are both near its center and the Vaal River meanders through it.

References

Corner, B. and Reimold, W. U., 1986, Aeromagnetic and gravity interpretation of the southern portion of the Kaapvaal Craton with special reference to the relationship between the Witwatersrand basin and the Vredefort dome *Meteoritics*, *21*, 347–48.

Hargraves, R. B., 1961, Shatter cones in the rocks of the Vredefort ring. *Proc. Geol. Soc. S. Africa*, *64*, 147–161.

Martini, J. 1978, Coesite and stishovite in the Vredefort dome, South Africa. *Nature*, *272*, 715–717.

8

Impact Structures of Asia

Beyenchime-Salaatin

Russia

Lat/Long: N71° 50′, E123° 30′
Diameter: 7.5 km
Age: <65 Ma
Condition: Eroded

The Beyenchime-Salaatin structure consists of a basin in sedimentary rocks, about 100 m deep at the center. There are outcroppings of impact breccias in the basin, partly covered by drift. The rim of the basin consists of deformed and uplifted beds, with evidence of the presence of overturned beds along a system of thrusts and radial joints. Shatter cones have been recovered from the shocked Proterozoic dolomites that are present.

Access: The Beyenchime-Salaatin structure is in north-central Siberia, about 100 km west of the Lena River and 200 km south of its delta. It is approximately at the tree line.

Reference

Masaitis, V. L., 1976, Astroblemes in the USSR. *Int. Geol. Rev.*, *18*, 1249–1258.

Bigach

Kazakhstan

Lat/Long: N48° 30′, E82° 00′
Diameter: 7 km
Age: 6 Ma
Condition: Eroded, partly exposed

Bigach is named for the settlement just to its north. It is discerned by its nearly circular shape, its relatively flat floor, and its rim, which is about 50 m high. The surroundings are hilly and semi-arid and the country rock is crystalline. The basin is cut by streams that drain to the east and the entire crater is filled with post-impact sediments. Farms take advantage of the relatively flat land in the crater floor. Crater ejecta can be recovered from as far away as 10 km from the crater, though most of it is covered by recent deposits.

Access: The structure is about 5 km south of the village of Bigach and about 40 km southwest of the town of Kokpekty.

Reference

Grieve, R. A. F., Wood, C. A., Garvin, J. B., McLaughlin, G. and McHone, J. F., 1988, Astronaut's guide to terrestrial impact craters. *LPI Technical Report*, 88–03, 89 pp.

Chiyli

Kazakhstan

Lat/Long: N49° 10′, E57° 51′
Diameter: 3 km
Age: 46 Ma

The small crater of Chiyli is located about 200 km south of Orsk.

Reference

Grieve, R. A. F., 1991, Terrestrial impact: the record in the rocks. *Meteoritics*, *26*, 175–194.

El'gygytgyn

Russia

Lat/Long: N67° 30′, E172° 05′
Diameter: 18 km
Age: 3.5 Ma
Condition: Somewhat eroded, partly lake-filled

El'gygytgyn is a beautifully-circular depression in the hilly, swampy tundra of eastern Siberia. It is partly filled by a square-shaped lake 160 m deep. The rim consists of a dissected ring of hills, reaching a height of about 450 m above the lake level. A river drains the area from the southeast, connecting to the Belaya River and the Bering Sea.

The surroundings of the crater are made up of volcanic formations from the Upper Cretaceous, overlain by Quaternary deposits in places. The rocks have a nearly horizontal attitude. The crater rim has been eroded to the point that only in a few places does the present edge of the rim correspond to the original lip of the crater. Radial and annular faults are present.

EL'GYGYTGYN: View from space of El'gygytgyn in winter (courtesy of NASA).

Curiously, no impact breccia has been reported from El'gygytgyn, though impact melts are found. These are mostly found in lake and stream deposits, and from three locales among the hills making up the rim. They consist of both solid glass bombs and frothy glass fragments. The largest glass bomb is a projectile 2 m long. The bombs typically show aerodynamic shapes, with evidence of twisting and turning in flight. They are covered with a typical 'bread-crust' cracking pattern.

Rocks and minerals from the crater show evidence of shock effects, ranging from planar features in quartz grains to the presence of coesite and stishovite. Some impactites contain lechatelierite, as well.

Because of its great depth and its isolation, the crater of El'gygytgyn is of interest to ichthyologists, as well as astronomers and geologists. In the 1970s and 1980s some new species of salmonid fishes (chars) were found there and in 1985 a fish belonging to a new genus was discovered, popularly called 'longfin char' and looking like an over-adorned salmon. It spends most of its time at great depth, only coming up near the warmer waters near the outlets of streams to digest its food. Study of specimens in 1990 established that the longfin char is the most primitive of the chars, apparently preserved in isolation in the frigid waters of this remarkable crater.

Access: El'gygytgyn is very remote and the accepted mode of travel seems to be by helicopter. The nearest city is

Magadan, 1400 km to the southwest. Take warm clothing; a recent scientific expedition there experienced eight snow storms in August.

References

Feldman, V. I., Granovskiy, L. B., Kapustkina, I. G., Karotayeva, N. N., Sazonova, L. V. and Dabizha, A. I., 1981, The El'gygytgyn meteor crater, in *Impactites*, ed. A. A. Marakushev, pp. 70–92. Moscow University.

Skopets, M., 1992, Secrets of Siberia's white lake. *Nat. Hist., Nov.*, 2–4.

Kara

Russia

Lat/Long: N69° 05′, E64° 18′
Diameter: 65 km
Age: 73 Ma
Condition: Partially buried

The Kara structure is a pronounced depression in the foothills of the Pay-Khoy Range. It is largely buried beneath about 100 m of Quaternary deposits. The country rock is made up of folded Paleozoic sediments. Extensive drilling and geophysical investigations have revealed characteristics typical of a large impact crater. The Ust-Kara structure, of similar age, lies adjacent to it.

Kara has a central uplift about 10 km across. Its rocks consist of highly-deformed examples of country rock with the orientations of the beds almost vertical. The outcrops show the presence of impact breccias and suevites. Minerals show the presence of shock metamorphism in the form of shock lamellae and planar cracks. Shatter cones are found in some of the rocks, especially the sandstones and siltstones.

Beneath the central uplift is the crater 'funnel', which is about 20 km in diameter. Cores show that it is filled in with breccias, suevites, glasses and related impact-generated rocks of large scale. The shape of the 'funnel' is somewhat asymmetric and steep-sided. It is surrounded by a zone of rim-like higher ground, where cuts made by stream valleys reveal the presence of more breccia and suevite.

The outer parts of the structure are characterized by shattered and pulverized country rock, largely revealed in the valleys of the Kara and other nearby rivers.

Access: The Kara depression lies on the shore of the Kara Sea near the mouth of the Kara River in a fairly remote part of northern Russia.

References

Koeberl, C., Sharpton, V., Murali, A. V. & Burke, K., 1990, Kara and Ust-Kara impact structures (USSR) and their relevance to the K/T boundary event. *Geology*, *18*, 50–53.

Masaitis, V. L., 1976, Astroblemes in the USSR. *Int. Geol. Rev.*, *18*, 1249–1258.

Sazonova, L. V., Karotayeva, N. N., Ponomarev, G. Y. and Dabisha, A. I., 1981, The Kara meteor crater, in *Impactites*, ed. A. A. Marakushev pp. 93–136. Izdatelstvo Moscow University.

Kar-Kul

Tajikistan

Lat/Long: N38° 57′, E73° 24′
Diameter: 45 km
Age: <225 Ma

Kar-Kul (or Karakul) is a large mountain lake about 30 km from the Chinese border.

Reference

Grieve, R. A. F., 1991, Terrestrial impact: the record in the rocks. *Meteoritics*, *26*, 175–194.

Logancha

Russia

Lat/Long: N65° 30′, E95° 48′
Diameter: 20 km
Age: 50 Ma
Condition: Highly eroded

The Logancha structure consists of about three-fourths of a circle with one sector (to the southwest) washed away by stream erosion. Its floor is less rugged than the surrounding hills, which are cut by many canyons and valleys. The Vivi River is to the south and the Tembechi River to the north. The crater rim is partly preserved and rises about 200 m above the floor. The target rocks are basalts and sedimentary. The presence of a basaltic target makes Logancha of importance for comparisons with lunar mare craters.

Access: Logancha is in a fairly sparsely-settled part of Siberia. The town of Tembenchi (or Tembechi) is about 100 km to the southeast.

Reference

Grieve, R. A. F., Wood, C. A., Garvin, J. B., McLaughlin, G. and McHone, J. F., 1988, Astronaut's guide to terrestrial impact craters. *LPI Technical Report*, 88–03, 89 pp.

Lonar
India

Lat/Long: N19° 58′, E76° 31′
Diameter: 1.8 km
Age: 0.05 Ma
Condition: Fairly fresh, partly lake-filled

Lonar Lake occupies a very special meteorite crater. The Lonar crater is the only fresh terrestrial crater that formed in basalt and is, therefore, important to studies of lunar craters, many of which, of course, have formed in basalt. The crater is 150 m deep and its rim rises about 20 m above the surrounding terrain. The central lake is shallow and salty.

Although the Lonar crater is obviously young, until the 1970s most geologists considered it to be a volcanic feature, in spite of the fact that the volcanic activity that formed the huge basaltic flows in this part of India ceased about 60 million years ago. Recent geological mapping, core drilling, and geophysical studies have now established its meteoritic impact origin decisively.

The target rocks consist of a series of basaltic flows that lie nearly perfectly-horizontally. At the crater rim these beds have been tilted upward, reaching to the vertical right at the rim crest, where in places they are even overturned. Under the lake sediments are layers of coarse impact breccia and microbreccia. The former is unshocked or mildly-shocked, while the microbreccia has grains that show various symptoms of shock metamorphism. The ejecta at the rim and beyond it also show two kinds of breccia: both throw-out breccia (bulky, unshocked fragments) and fallout breccia (an overlying layer including shocked rocks).

LONAR: Lonar Lake from the eastern rim, with ejecta visible in the foreground (courtesy of K. Fredriksson).

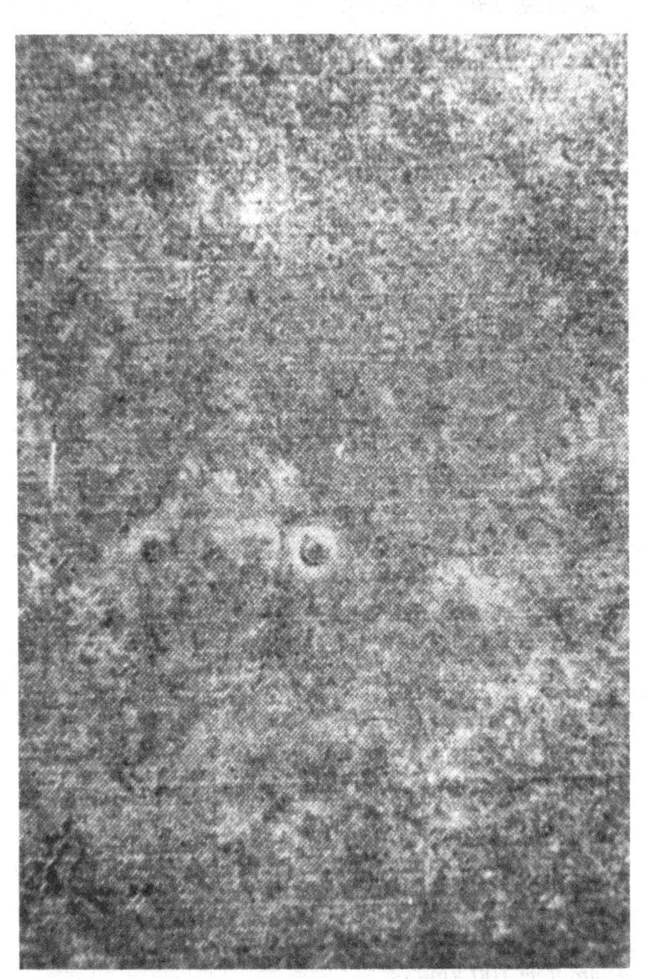

LONAR: View from space of the Lonar crater (courtesy of NASA).

Dark glass spherules and droplets are recovered from the mixed breccia and these show both flow structure and shock effects. They are similar to microtectites. Larger pieces of glass resemble the 'Fladlen' of the Ries structure.

A small (300 m diameter) depression to the north of the rim has been suggested as a possible second crater.

Access: The crater is easy to visit. The town of Lonar lies on its northeast flank and there is a guest house on its rim. Lonar is about 350 km northwest of Hyderabad and 400 km east-northeast of Bombay.

Reference

Fredriksson, K., Dube, A., Milton, D. J. and Balasundaram, M., 1973, Lonar Lake, India: an impact crater in basalt. *Science, 180*, 862–864.

Macha

Russia

Lat/Long: N59° 59′, E118° 00′
Diameter: 0.3 km
Age: <0.007 Ma
Condition: Fresh

The Macha crater is located in central Siberia near the Lena River and not far from the village of Kochegarovo.

Reference

Grieve, R. A. F., 1991, Terrestrial impact: the record in the rocks. *Meteoritics, 26*, 175–194.

Popigai

Russia

Lat/Long: N71° 30′, E111° 00′
Diameter: 100 km
Age: 35 Ma
Condition: Somewhat eroded and partially filled with sediments

The Popigai crater is Russia's largest and the youngest of the Earth's 100-km-class craters. Considering its size and age, it is well-preserved, with much of its crater topography evident. There is a 300-m-deep crater, some 70 km in diameter, partially filled by Quaternary conglomerates and sands. The location is on the northern boundary of the Anabar Shield, where dolomites, quartzites, sandstones, and siltstones overlie the basement gneisses. The flat floor of the crater is surrounded by spectacular cliffs made up of impact breccias, which include giant blocks of country rock. Below the present surface is a basin with a central uplift about 45 km in diameter and rising 2 km above its surroundings. The uplift is made up of gneissic breccias and it is surrounded by a ring of suevites, which are laced by inclusions of massive impactites of an unusual nature, called 'tagamites' (named after the Tagamy Ridge in Popigai crater). The suevite layer is some 2 km thick.

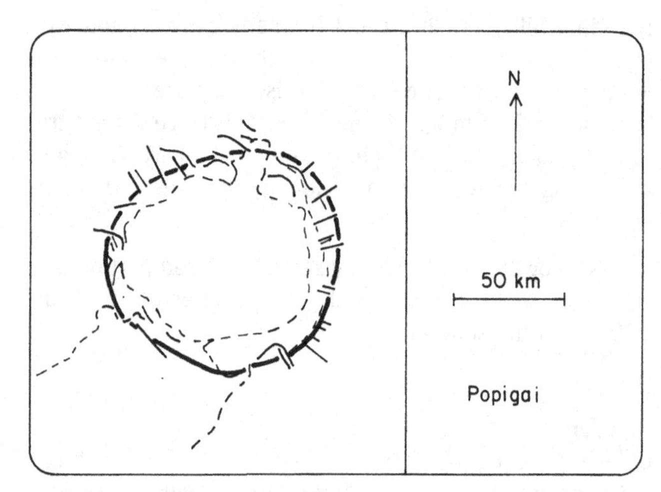

POPIGAI: Geological sketch map of the Popigai crater (after Masaitis).

There is a flat, asymmetric outer rim about 25 km wide. A 150-m-thick layer of blocky breccia surrounds the inner parts of the crater. It, in turn, is surrounded by an outer area of intensely-deformed sedimentary rocks. The structure of the outer rim is unusual: there are tongues cut into the edges of the rim, filled with breccia and impactites, and radial faults as well as concentric synclines and anticlines are present.

Access: The Popigai crater is in north-central Siberia, just above the tree line and about 400 km south of the Laptev Sea. It lies near the headwaters of the Popigai River.

Reference

Masaitis, V. L., 1976, Astroblemes in the USSR. *Int. Geol. Rev., 18*, 1249–1258.

Ragozinka

Russia

Lat/Long: N58° 18′, E62° 00′
Diameter: 9 km
Age: 55 Ma

The Ragozinka structure is just east of the Urals, near the Tura River.

Reference

Grieve, R. A. F., 1991, Terrestrial impact: the record in the rocks. *Meteoritics*, *26*, 175–194.

Shunak

Kazakhstan

Lat/Long: N47° 12′, E72° 42′
Diameter: 3.1 km
Age: 12 Ma
Condition: Exposed, eroded

Shunak is a circular crater with a raised rim. The rim is cut in the southeast by the 200-m-wide valley of an outlet stream, which flows intermittently. The inner edges of the rim are concave, with an average slope of about 25°, while the outer slopes are more gentle. Radial stream valleys dissect the rim slopes and dry stream channels on the crater floor point towards the outlet breach of the rim. The rocks at the crest of the rim show typical impact-crater orientations: they are upturned and twisted.

Geophysical study has shown that the crater floor is about 600 m below the rim, with about 200 m of that filled in by lake deposits. Below these post-impact deposits are breccias made up of the various volcanic rock types found in the Devonian target rocks. One of the boreholes brought up breccias that included fragments of light brown glass, found to be impact melt glass. Shatter cones are found and shock features show up in quartz grains in the breccia.

Access: The crater is in an arid, sparsely-populated area northwest of Lake Balkhash. It is 40 km west of the railway station of Mointy and 160 km west-northwest of the city of Balkhash.

Reference

Feldman, V. I., Granovskiy, L. B. and Dabisha, A. I., 1981, The Shunak meteor crater, in *Impactites*, ed. A. A. Marakushev, pp. 56–69. Moscow University.

Sikhote Alin

Russia

Lat/Long: N46° 07′, E134° 40′
Diameter: 0.027 km
Age: Formed in 1947
Condition: Fresh

The craters of the Sikhote Alin meteorite shower were formed at a little past nine o'clock in the morning, local time, on 12 February, 1947. A small iron asteroid collided with the Earth's atmosphere at a velocity of 14.5 km/s and broke up into many pieces at an altitude of about 5 km. These pieces then hit the ground in the forest of the western part of the Sikhote Alin mountain range in southeastern Siberia. They created 122 craters larger than 0.5 m and many smaller holes, with a total of 383 places of impact. The meteorites were contained within an elliptical area of 1.6 km^2.

The craters at Sikhote Alin provide an important, uneroded encyclopedia of cratering effects. The largest crater is small, 27 m in diameter, but it and the other 121 craters clearly demonstrate how the range in characteristics of small craters depends on the size of the projectile. For the largest craters, the meteorite was destroyed by the impact and its fragments were distributed about the floor

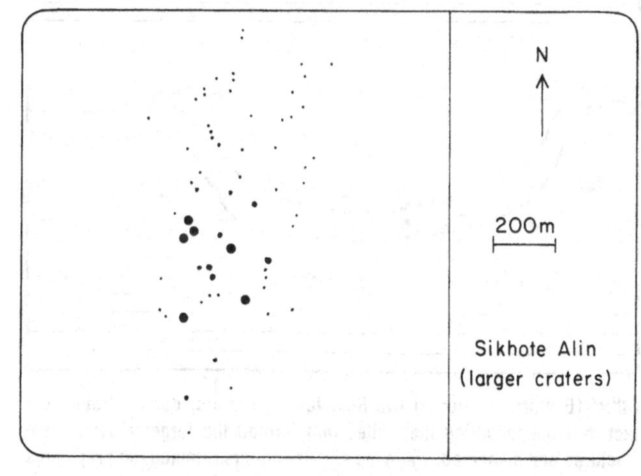

SIKHOTE ALIN: Map of the locations of the larger craters at Sikhote Alin.

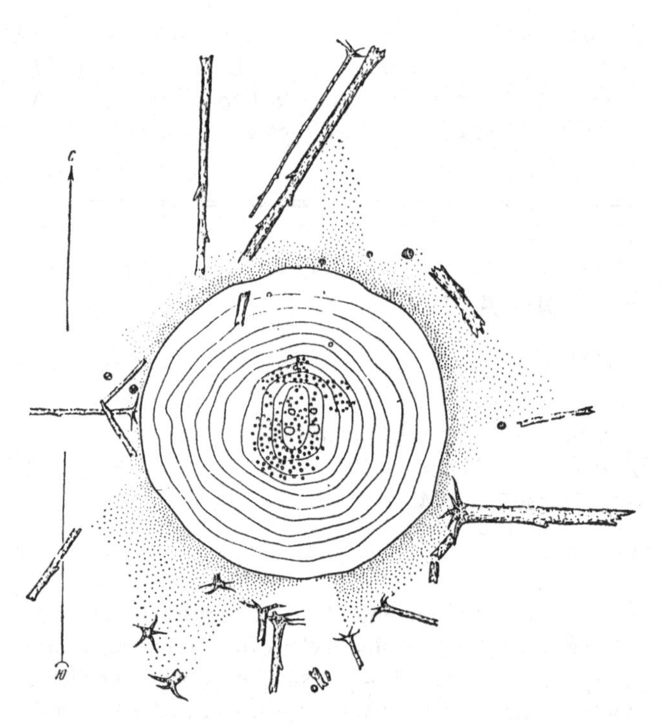

SIKHOTE ALIN: Map of Sikhote Alin crater 10, showing the blown-down trees and the locations of meteorite fragments in the crater bottom (from E. L. Krinov).

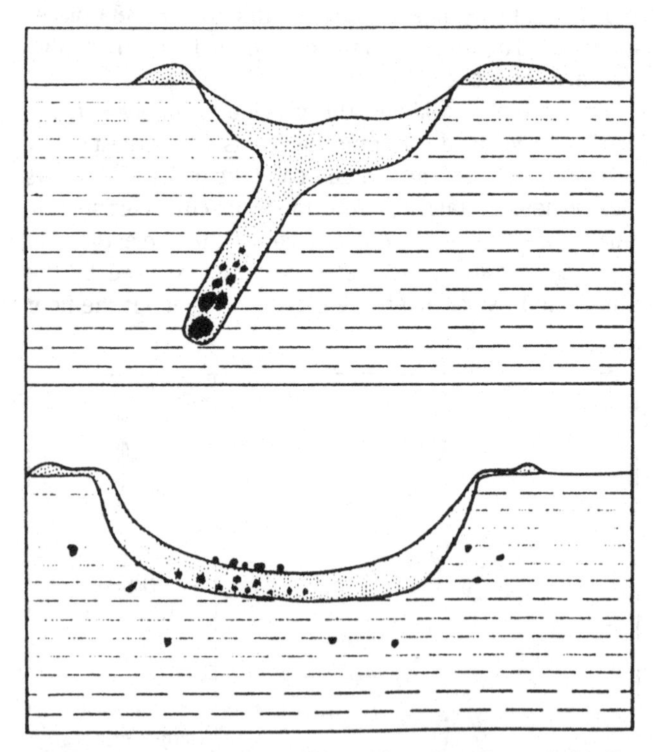

SIKHOTE ALIN: Profiles of two Sikhote Alin craters, demonstrating the fact that the incoming meteorites that formed the larger craters were fractured and dispersed, while the smaller craters retained the projectiles intact (from E. L. Krinov).

SIKHOTE ALIN: A painting of the Sikhote Alin meteorite as it was witnessed by an artist in a nearby town (from a postage stamp of the USSR).

SIKHOTE ALIN: Among the Sikhote Alin craters, Russian meteoriticists discover the first specimen of iron meteorite (courtesy of E. L. Krinov, who is at the right, holding the meteorite).

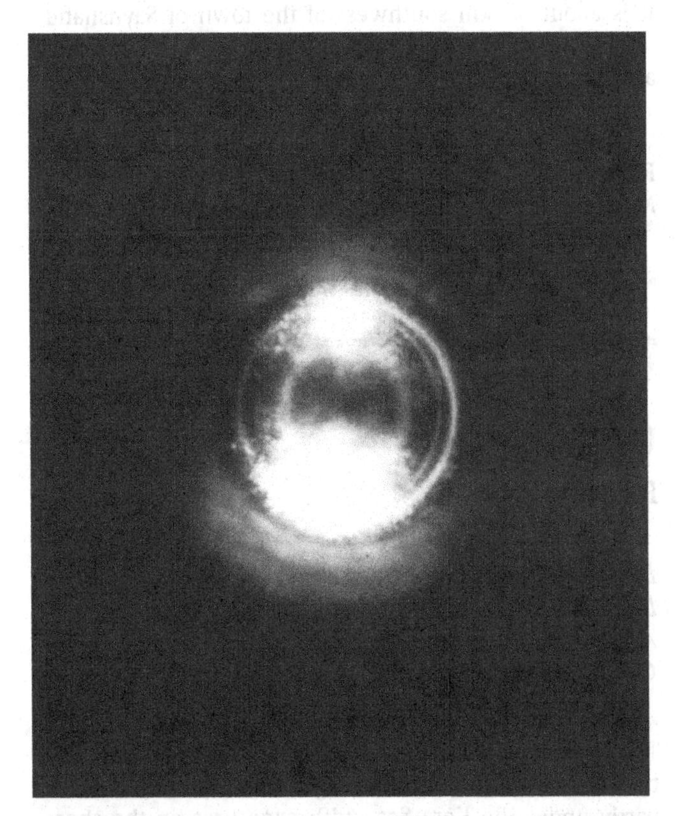

SIKHOTE ALIN: Photomicrograph of a Sikhote Alin ablation spherule, collected at the impact site.

of the crater and its surroundings. For the smaller craters, the meteorite usually remained intact, having dug a hole into the ground beneath the crater. The channels so dug ranged up to 8 m in length and were directed to the south (the meteorites were witnessed to come in from the north). For some of the smaller craters, the meteorite bounced back out, landing some distance away.

More than 300 eyewitnesses reported on the circumstances of the fall of the meteorites. From their descriptions, an orbit for the projectile was calculated, and it turned out to be that of an asteroid with its aphelion in the asteroid belt. The witnesses reported that the fireball was so bright that it hurt their eyes and it formed secondary shadows of objects in addition to the sun's shadows. When the fireball had disappeared, loud, explosive sounds were heard, followed by a roaring noise. Doors were blown open, windows were broken, and plaster fell from ceilings. A huge dust train remained in the sky to mark the flight of the meteorites for several hours.

Scientific expeditions to the place of fall were organized quickly, and a series of intensive studies, completed in 1950, provided the most comprehensive and well-documented account of a meteoritic event ever made. Several books and extensive published reports provide complete data on all of the craters, all of the meteorites and all of the attendant disturbance of the ground and forest. Examples of the latter include cases where trees were uprooted around craters, leaving them as a radial outer ring and cases where trees had their tops broken off and where holes were dug into trees. The 'ellipse of fall' is particularly well-documented, with the smaller, unbroken meteorites being found at the northern end of the ellipse and the larger craters at the south end.

Most of the craters were excavated and their meteorites removed. However, three craters of a range in size were left untouched, with protective roofs built over them so that they will be preserved as natural monuments.

Access: The craters are located in the western foothills of the Sikhote Alin mountains, east of the town of Guberovo, which is on the Trans-Siberian railroad line. The nearest larger city is Iman, which is also known as Dal'nerechensk, and which is about 50 km southwest of the crater field.

References

Krinov, E. L., 1963, The Tunguska and Sikhote-Alin meteorites, in *The Solar System*, vol. 4, *Moon, Meteorites and Craters*, ed. B. Middlehurst and G. Kuiper, pp. 208–234. University of Chicago Press.

Krinov, E. L., 1966, *Giant Meteorites*. Pergamon Press, New York, 397 pp.

Sobolev

Russia

Lat/Long: N46° 18′, E138° 52′
Diameter: 0.05 km
Age: 200 years
Condition: Fresh

The Sobolev crater is one of the youngest meteorite craters known. It age has been estimated by analysis of the degree of weathering of the ejecta and of the condition of the ejecta-covered target soil. It is located on a wooded slope adjacent to the Gorelyy Creek and is asymmetrical, with its eastern rim much lower than its western, because of erosion towards the creek bed on the east. The target rock is a mixture of Cretaceous and Tertiary volcanics. Fragments of these rocks, making up a loose breccia, extend out to distances of 50 m from the crater. The ejecta blanket overlies a thin band of clay that was the original pre-impact soil. This soil contains remnants of roots and branches. Microscopic meteoritic and impact melt material has been recovered from the breccia in the form of spherules with enhanced nickel content, irregular particles with a glazed silicate covering, and silicate spherules.

Access: The Sobolev crater is in the Sikhote Alin Mountains east of the Sikhote Alin craters. It is very near the coast and about 600 km north-northwest of Vladivostok.

Reference

Khryanina, L. P., 1981, Sobolevskiy meteorite crater (Sikhote-Alin range). *Int. Geol. Rev.*, *23*, 1–10.

Tabun-Khara-Obo

Mongolia

Lat/Long: N44° 06′, E109° 36′
Diameter: 1.3 km
Age: >120 Ma
Condition: Somewhat eroded, sand-filled

The Tabun-Khara-Obo crater is located in the Gobi Desert. It consists of a circular, flat basin with a rim that rises 20 to 50 m above the floor. The structure interrupts a northeast–southwest striking sedimentary massif that is deeply etched by prevailing winds.

Access: The structure is in a remote part of the eastern Gobi Desert, about 470 km south-southeast of Ulaan Baatar. It is about 80 km southwest of the town of Saynshand, which is connected by road and train with Ulaan Baatar and China.

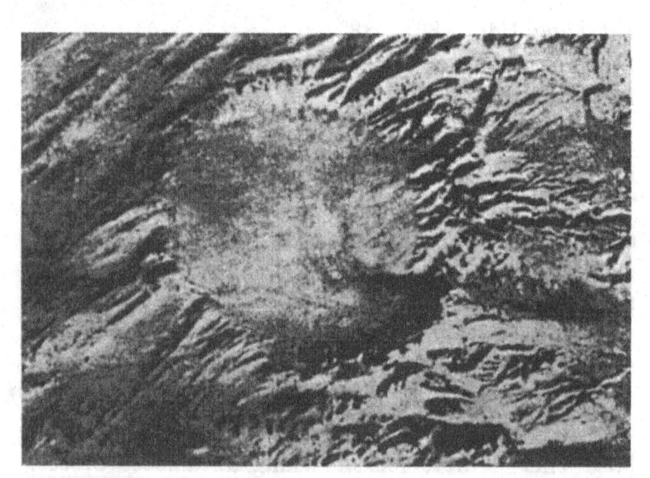

TABUN-KHARA-OBO: Aerial photograph of the Tabun-Khara-Obo structure (from Suetenko and Shkerin).

Reference

McHone, J. F. and Dietz, R. S., 1976, Tabun Khara Obo crater, Mongolia: probably meteoritic. *Meteoritics*, *11*, 332–334.

Ust-Kara

Russia

Lat/Long: N 69° 18′, E65° 18′
Diameter: 25 km
Age: 73 Ma
Condition: Under water

The Ust-Kara structure is northeast of the Kara crater. The two craters are of the same age. The Ust-Kara crater is partly under the Kara Sea, with exposures on the shore near the Kara River estuary.

References

Grieve, R. A. F., 1991, Terrestrial impact: the record in the rocks. *Meteoritics*, *26*, 175–194.
Koeberl, C., Sharpton, V., Murali, A. V. and Burke, K., 1990, Kara and Ust-Kara impact structures (USSR) and their relevance to the K/T boundary event. *Geology*, *18*, 50–53.

Wabar

Saudi Arabia

Lat/Long: N21° 30′, E50° 28′
Diameter: 0.1 km
Age: 0.006 Ma
Condition: Fresh, covered with drifting sand

The discovery of the Wabar craters was the climax of a romantic and mysterious quest. Legends told of a great and ancient city in the 'Empty Quarter' of Arabia, a city

WABAR: The Wabar crater, largely filled with sand (from Philby).

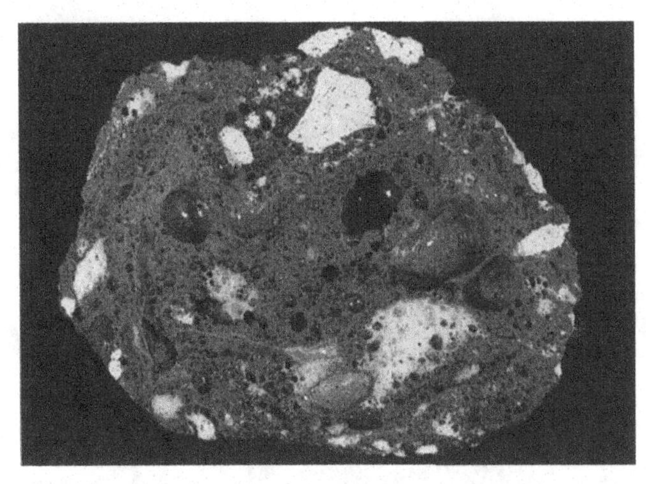

WABAR: Frothy Wabar glass is found around the craters. Except for the presence of iron–nickel spherules, its chemistry is identical to that of the surrounding sands.

where riches abounded and licentious behavior ruled. It was destroyed by an angry God, who smote it with a violent stroke from the heavens. Now, the legend went, only low ruins of the mansions' walls and a curious iron statue the size of a camel remained. The story is recounted in the book, '*The Empty Quarter*' by H. St. J. Philby, who set out in 1932 to find the fabled ruins. Conflicting accounts of its location and its name added suspense to the expedition, but one of the Arab guides knew the location well enough that the party found it late one evening. Instead of a ruined city, they found the walls of two meteorite craters. Instead of a camel-sized iron statue, they found a rabbit-sized iron meteorite. In the following days they found suggestions that other craters, perhaps three others, lay hidden under the shifting sands, and found more meteorite fragments, as well as large quantities of gray, frothy impact glass. Many small, black glass beads, the 'Pearls of al'Ad', were the meteoritic substitute for the promised jewels of the lost city of Wabar.

Not very much is known about the geology of the Wabar craters. Authorities even disagree on how many craters there are. Although Philby found two, large well-exposed craters, an aerial photograph made in 1961 by D. Holm recorded only one, others having been covered by wind-blown sand. Our knowledge about the impact has been derived primarily from the Wabar glass and the meteorites. The glass contains many iron–nickel spherules derived from the melted meteoritic component and is otherwise similar in composition to the surrounding sands. Coesite has been identified, confirming the impact origin of the material. The meteorite that caused the craters was a medium octahedrite.

Access: Traditionally, Wabar (also known as Al Hadidah) was approached by camel train. Now it is more readily reached by an appropriate overland vehicle, although there are no roads. Wabar is 500 km southeast of Riyadh.

References

Philby, H. St. J., 1933, *The Empty Quarter*. Holt and Co., New York.

See, T. H., Horz, F. and Murali, A. V., 1988, Two types of impact melt from Wabar crater, Saudi Arabia. *Lunar Plane. Sci.*, *19*, 1053–1954.

Zhamanshin

Kazakhstan

Lat/Long: N48° 24′, E60° 58′
Diameter: 13.5 km
Age: 0.9 Ma
Condition: Exposed and eroded

The Zhamanshin crater has a central basin that is partly filled with lake deposits, an inner rim that is circular and eroded, especially in the east, and an outer ring. The most unusual feature of the crater is its wide variety of impact glasses, also well-preserved. It is one of the few craters associated with tektites (the Irghizites, named after the nearby Irghiz River) and it also has several interesting kinds of impact glass, one of which is similar to suevite and is referred to as Zhamanshinite, described below.

The target rocks are Cretaceous and Tertiary sedimentary rocks that overlie a stable Paleozoic folded basement.

ZHAMANSHIN: The southeast part of the rim of the Zhamanshin crater. The white hills are composed of Paleogenic soft loams and clays and the dark hills (left) are Paleozoic rock fragments (courtesy of E. Izokh).

ZHAMANSHIN: A large 'bomb' of impact melt from the Zhamanshin crater, 30 cm long (courtesy of E. Izokh).

ZHAMANSHIN: The south rim of the Zhamanshin crater rim, looking northeast. Meteoriticists are seen examining the Zhamanshinite strewn field. Beyond the small hill, which consists of allogenic impact breccia, the central crater plain is visible, with the north rim making up the distant horizon (courtesy of E. Izokh).

ZHAMANSHIN: A piece of impactite from the north rim of the Zhamanshin crater. Saw-cut interior surface, showing fluidization (courtesy of E. Izokh).

The sedimentary rock layer is about 150 m thick. The inner crater basin is about 6 km in diameter and there is an outer rim that lies out at twice that distance. It consists of hills with an average relief of about 50 m, made up primarily of impact breccia. The rim is broken in the east by a low place, called 'the gate', which drains the crater. A core drilled at the center of the basin shows the presence of 55 m of post-impact sediments.

There is abundant evidence of shock effects at Zhamanshin. Shatter cones have been recovered from the southeast rim and breccias show the presence of coesite and stishovite. Maskelynite, derived from feldspar, is also present.

The Irghizites are very similar to tektites, such as the Australites and the Indochinites, but differ from those in being restricted to a small geographical area, the vicinity of the crater. Thus they are probably intermediate between tektites and impact glass in their mode of formation, a situation that is also supported by their chemical composition. Two forms of Irghizites have been recovered: small millimeter- to centimeter-sized irregular, ropy pieces of black glass, and micro-irghizites, which are smaller and which range in color from yellow to black and in shape from spherules to tear-drops to irregular fragments. The micro-irghizites, which were collected from sediments in a stream bed, show a wide variety of chemical composition, which is interpreted as evidence for a mix of target materials, probably a soil derived from sedimentary rocks rich in silicate. Traces of iron-nickel suggest the presence of meteoritic material in the micro-irghizites, indicating that they are not strictly tektites, but were formed after the projectile had begun to evaporate and had mixed with the melted and evaporated target rocks.

Zhamanshinites can be separated into several types, which differ most notably in their silicate content. The silicate-rich zhamanshinites have about 75% SiO_2, while the silicate-poor varieties have as little as 40% SiO_2. They are large, often blocky fragments of glass, black or dark green in color and often showing flow structures, bubbles and inclusions. Many examples look like Ries suevite, except for their color. While any individual irghizite will tend to be homogeneous in chemical composition, each zhamanshinite can show a wide variety of composition. These objects are examples of impact glass that was formed low in the impact event profile and was either left in the crater or ejected to limited distances from the crater center.

Access: The Zhamanshin crater has been visited by several scientific expeditions, including international ones. It is 100 km north of the Aral Sea and 40 km southwest of the village of Irghiz.

References

Bouska, V., Povondra, P., Florenskij, P. and Randa, Z., 1981, Irghizites and zhamanshites: Zhamansin crater, USSR. *Meteoritics*, *16*, 171–184.

Izokh, E., 1991, Zhamanshin impact crater and tektite problems. *Geol. Geophys.*, *4*, 5–14.

Masaitis, V. L., Boiko, Y. I. and Izokh, E., 1984, Zhamanshin impact crater (western Kazakhstan): additional geological data. *Lunar Planet. Sci.*, *15*, 515–516.

Index